Research

Ethics

Research Ethics

A READER

EDITED BY

Deni Elliott, Ed.D. and Judy E. Stern, Ph.D.

PUBLISHED BY UNIVERSITY PRESS OF NEW ENGLAND

for the Institute for the Study of Applied
and Professional Ethics at Dartmouth College

Hanover and London

Published by University Press of New England,
One Court Street, Lebanon, NH 03766
www.upne.com

Printed in the United States of America 5 4 3 2

CIP data appear at the end of the book

ISBN–13: 978–0–87451–797–2
ISBN–10: 0–87451–797–4

Contents

Introduction

In 1991, Dartmouth College, like other research institutions around the country, was struggling with the need to deal with increasingly strict federal requirements to include ethics in the research training of its students of science and engineering. Unlike many institutions at the time, however, Dartmouth had an established Institute for the Study of Applied and Professional Ethics. This Institute, dedicated to interdisciplinary research and teaching, was supported by a consortium of faculty members, more than 200 strong, that bridged Dartmouth's professional schools of business, engineering, and medicine along with the liberal arts college and graduate school. A University Seminar series focused the need for Dartmouth faculty, along with science and engineering faculty around the country, to come to terms with the increasingly stringent Federal mandate to teach ethics as part of research training.

Two projects developed out of those discussions. The first, funded by the Department of Education's Fund for the Improvement of Post-Secondary Education, September 1992–August 1995, allowed a small group of Dartmouth scientists and philosophers to train faculty in teaching research ethics, develop a pilot graduate-level course in research ethics for students of science and engineering, and develop methods for evaluating the success of the course. The second project, funded by The National Science Foundation, brought together a national consortium of philosophers, scientists, and engineers for the purpose of developing materials for the teaching of research ethics in science and engineering. The projects had overlapping researchers: Deni Elliott, the principal investigator on the NSF project was also co-director on the FIPSE project. Dartmouth philosopher Bernard Gert and scientist Edward Berger were members of the FIPSE Dartmouth-based research group and were also members of the NSF national consortium. It's natural that a shared product would develop from the projects. That product is this volume, edited by the co-directors of the FIPSE project, Deni Elliott and Judy E. Stern.

This reader reflects the topics covered in the course in research ethics for graduate students of science and engineering taught at Dartmouth in 1993 and 1994. The primary topic essays for most of this book's chapters are edited from longer works produced by the NSF consortium members; many of the case studies included for discussion were developed through the NSF-supported work.

However, the grants took slightly different directions, reflecting the makeup of each group. The FIPSE course spent four seminar sessions on the topics of methodology, reporting research, peer review, and funding. The NSF consortium produced a single piece that spanned all of these topics. The FIPSE course did not include specific work on conflicts of interest; the NSF materials include an entire section devoted to this issue. The FIPSE course also included sections on experimentation involving animals and humans, not addressed by the NSF consortium.

This volume is designed as a student reader to bring together original material and articles reprinted from other sources that resulted from the NSF and FIPSE projects. The original articles lay the framework for each topic area. The reprinted articles add interesting case material and some additional perspectives for student evaluation. In addition, specific cases for discussion, some in narrative form and some in dialogue form, follow each chapter. It is not our intention that every course use the entire book cover-to-cover. Instead, we expect that faculty and students, like our separate but complementary project groups, will use materials contained here in a way that meets the special needs and interests of their researchers and institution.

More in-depth treatment of the topics covered by the members of the NSF consortium and additional cases for discussion can be found in Vol. 4, Nos. 3 and 4 of *Professional Ethics: A Multidisciplinary Journal,* which has devoted a special double issue to the consortium's work. The topic areas covered in the reader are outlined below.

Chapter 1, Teaching and Learning Research Ethics, by NSF consortium members Judith Swazey and Stephanie Bird, developed through the working group's realization that students and teachers in departments of science and engineering are often puzzled about what it means to teach ethics and why it should be taught.

Chapter 2, Morality and Scientific Research, by Bernard Gert, developed through his work on both projects. This chapter is also the result of a high level of commitment by scientists and engineers at Dartmouth and from across the country. Gert's understanding of scientific convention and the reality of life in the laboratory developed through many hours of sometimes patient and sometimes heated conversation.

This chapter is followed by several reprinted articles that bring up a number of cases for discussion and consideration. The cases are introduced in chapter 3 by Allan U. Munck, a member of the FIPSE project faculty, who offers insight into research misconduct from the perspective of an experienced laboratory scientist. These cases can be referred to while discussing the issues raised throughout this volume. The first article by Racker presents several clearcut cases of misconduct. It is followed by several articles dealing with central features of the highly publicized Imanishi Kari/Baltimore case. The final article in the section describes a very complicated case that occurred at Cal Tech and brings up a variety of issues, from methodology and reporting to institutional responsibility.

Chapter 4, Relationships in Laboratories and Research Communities, by NSF consortium members Vivian Weil and Robert Arzbaecher, grew out of the consortium's understanding that students often confront their first ethical problems in the relationships they share within the working group in the laboratory. The hierarchical power structure, with students often at the bottom rung, makes them vulnerable to exploitation. The kinds of responsibilities individuals have toward one another provide a natural avenue for students to begin grappling with ethical issues in research. The chapter presents a number of hypothetical cases for consideration. It is followed by a reprinted article describing a complicated case of interpersonal interactions gone awry. Because this case was also poorly handled at the institutional level, it is offered for consideration again at the end of the chapter on institutional responsibility.

Chapter 5, Conducting, Reporting, and Funding Research, by NSF consortium members Stephanie Bird and David Housman, leads the way through the morass of professional decisions made in the research process, from creating hypotheses through reporting results. The authors' analysis lays bare the value decisions that are implicit in the research process, raising issues in the conducting and reporting of research that are central to what students need to understand about research ethics. The chapter therefore introduces a number of related articles and cases that explore this issue in more depth.

Chapter 6, Conflicts of Interest and Conflicts of Commitment, by NSF consortium members Patricia Werhane and Jeffrey Doering, moves the issue of relationships outside the laboratory. It explores the researcher's need to protect self and society from situations in which self-interest competes with scientific judgment and discusses how researchers ought to juggle conflicting commitments within their profession—students, research, institutional commitments to the school, as well as institutional commitments to the profession and society—and the unavoidable conflict between personal and professional commitments. This chapter is also followed by a reprint of an interesting and complex case.

Chapter 7, Institutional Responsibility, by NSF consortium members Edward Berger and Bernard Gert, broadens the question of responsibility beyond the investigator and the individual laboratory to include the research institution as a whole. Beyond federal mandates, these authors explore what an institution ought to do to meet its moral responsibilities to investigators, to those who suspect misconduct or questionable practices, and to those who are accused. Three articles offering additional insight into the definitions of misconduct follow this chapter.

In Chapter 8, Animal Experimentation and Ethics, Deni Elliott and Marilyn Brown argue that the question of the moral standing of animals is irrelevant to the fact that researchers have obligations to treat animals with respect and care. This chapter is followed by additional information about the requirements involved in animal experimentation and by a thought-provoking article on the animal liberation movement.

Chapter 9, Human Experimentation, by Judy Stern and Karen Lomax, examines the dilemmas that arise from the need to experiment on human beings and the moral requirement that researchers protect human beings who are vulnerable to their studies. Emphasis is placed on the role and responsibilities of basic researchers who become involved in human subject research. The chapter is followed by a copy of the Nuremberg code and the Declaration of Helsinki, which are referred to in the chapter and with which all researchers should be familiar. An article elaborating on a group of cases involving radiation experiments on human subjects is also included.

A companion book to this coursebook, *The Ethics of Scientific Research: A Guidebook for Course Development,* also published by the University Press of New England, details the FIPSE-supported Dartmouth group's experience in training faculty, teaching, evaluating, and implementing the course at Dartmouth College. The extensive bibliography, divided by topic area, provides a beginning for additional research in these areas. The teaching book was published in the hope that others can learn from our successes and from our mistakes.

D.E. and J.E.S.

July 1996

Research

Ethics

1

Teaching and Learning Research Ethics

JUDITH P. SWAZEY and STEPHANIE J. BIRD[1]

I. INTRODUCTION

This chapter is written for those in the natural and physical sciences and engineering who are or may be engaged in teaching or learning about research ethics. Such teaching can take place in various contexts, uses a variety of pedagogical methods, and may be done solo or in an interdisciplinary team-teaching format. The chapter also has been written for students, because both faculty and students should think of themselves in the double role of teachers and learners, mutually engaged in analyzing, discussing, and dealing with the complex issues that are the subject matter of scientific research ethics.

There are many differences, as well as similarities, within and between the cultures and conventions of the disciplines and subdisciplines in science and engineering that affect the content of what is taught, as well as how and by whom it may be taught most effectively. Further, although we are dealing with research ethics only in the context of higher education in the United States, even here there is a great amount of sociocultural diversity among the students and postdoctoral fellows, faculty, administrators, and others involved in many fields of academic research. It is important to be aware of this diversity, and not to assume that everyone approaches the subject from a common social and cultural framework. People from different societies and cultural traditions have a range of attitudes, beliefs, and values that affect how they define ethically "acceptable" standards and practices, what they define as "ethical" issues, and how they perceive and respond to those issues.

With these caveats in mind, this chapter first addresses two questions that are often asked by those new to the domain of research ethics: Can ethics be taught to adults, and why teach research ethics?

We then discuss who should teach research ethics, various teaching contexts and formats, and pedagogical methods, focusing particularly on the use of case studies. This section includes an example of a framework that can be used for learning to identify and reason about ethical issues in research. Finally, we include a case study to illustrate the use of the framework.

II. CAN ETHICS BE TAUGHT TO ADULTS?

There are a number of barriers that have impeded the incorporation of explicit teaching about scientific research ethics into the education of graduate and undergraduate students. One such obstacle is the belief that by the time a person enters graduate or professional school, and perhaps even college, their values and ethical standards cannot be refined or altered. In the Acadia Institute's nationwide survey of science and engineering faculty, for example, over 40 percent of the respondents strongly agreed or agreed with the statement that "by the time students enter graduate school, their values and ethical standards are so firmly established that they are difficult to change."[2]

There is, however, a solid body of research to refute the belief that ethics cannot be taught to adults. In terms of our focus—teaching and learning scientific research ethics—some of the most relevant work has been conducted by educational psychologist James Rest and his colleagues at the University of Minnesota. The extensive research by Rest and others demonstrates that a person's moral development—the way the person approaches and resolves ethical issues—continues to change throughout formal education. Rest has summarized the following five major findings from moral development studies:

1. Dramatic and extensive changes occur in young adulthood (the 20s and 30s) in the basic problem-solving strategies used by the person in dealing with ethical issues. That is, the basic assumptions and perspectives by which people define what is morally right or wrong change in this period, and the change is just as dramatic and fundamental as changes in the years before puberty.

2. These changes are linked to fundamental reconceptualizations in how the person understands society and his/her stake in society. The changes are not merely attitudinal fluctuations that vary with passing fads and fancies.
3. Formal education (years in college/professional school) is a powerful and consistent correlate to this change. The general picture is that development continues as long as a person is in a formal educational setting; development plateaus when a person leaves school.
4. Deliberate educational attempts (formal curriculum) to influence awareness of moral problems and to influence the [moral] reasoning/judgment process can be demonstrated to be effective [by pre/post assessment tests] . . .
5. Studies link moral perception and moral judgment with actual, real-life behavior. Scores on psychological tests of internal processes (of moral sensitivity and of moral judgment) are significantly correlated with behavior. The linkages are of three kinds:
   (a) correlations of test scores with how ethical issues are dealt with in the professional setting;
   (b) correlations of test scores with ratings of general professional performance (without particular regard to ethical aspects of professional conduct, but with overall professional competence);
   (c) correlation of test scores with moral behavior in general (without regard to one's profession . . .).[3]

Thus, undergraduate and graduate students can learn moral awareness or sensitivity about ethical issues in scientific research. Further, they can be taught the skills that will enable them to reason about those issues and to formulate moral arguments and judgment processes for developing one or more acceptable solutions to difficult cases, as well as recognizing what constitutes unacceptable decisions and actions, and apply their knowledge and skills to their work as a scientist.

III. WHY TEACH ACADEMIC RESEARCH ETHICS?

Granted that ethics can be taught to adults, why teach academic research ethics? Graduate school deans, faculty, and graduate students have all stated that "ethical preparedness training"—preparing students to recognize and deal with ethical issues they may encounter in their field—should be an important function of their universities and departments.[4] Overall, 99 percent of the deans, 88 percent of the faculty, and 82 percent of the students who responded to these surveys said that their institutions (deans) or departments (faculty and students in the same departments) should take a very active to somewhat active role in such training.

However, only 49 percent of the deans who were surveyed in 1988 felt that their institution was actually doing a very or quite effective job in the educational area. When the faculty were surveyed in 1991, only 4 percent judged that their department was very active and 37 percent that it was somewhat active, while 13 percent said that it was not at all active. Graduate students, who were surveyed in 1990, gave their departments even lower marks than did faculty: Only 3 percent thought their department was very active and 25 percent reported that it was not at all active.

However important many people may think it is to include research ethics as part of graduate training in science and engineering, these findings demonstrate that there is a substantial gulf between stated importance and activity levels. Explicit training in research ethics is only slowly becoming a recognized part of the "tools" for doing scientific research. Reasons range from the pressures of time and crowded curricula to the view of some faculty that learning research ethics is not an integral part of doing science or being a scientifically "good" scientist. Others believe that research ethics is too far outside their area of expertise to contemplate including it in the education of their students. Physician and medical ethicist Bernard Lo has pithily summarized why some skeptical scientists argue that teaching research ethics is "unnecessary, unhelpful or even counterproductive." The skeptics offer the following types of declarations: "Only unethical persons have ethical problems; ethics is being a good person, not a system of rules; every case is unique, so guidelines are impossible; ethics is merely personal belief; studying ethics doesn't help solve real problems."[5]

Other reasons for faculty doubting the need for or effectiveness of teaching research ethics were suggested by the Acadia Institute survey. Most faculty respondents (74 percent) believe that all or at least a majority of the doctoral students in their department "exhibit an awareness of ethical standards and issues in their discipline."[6] The bases for this judgment need to be explored by programs working to introduce the teaching of research ethics. Other findings in the Acadia study indicate that professional values and ethics are discussed infrequently in most graduate training programs, and that many of the standards that do or should govern research practices in various disciplines are not clearly articulated.

Faculty also may be ambivalent about the importance of explicit research ethics teaching because they are unfamiliar with the substantive content of ethics and value studies. One red herring is the belief that examining the ethical dimensions of research standards and practices will diminish the scientist's/engineer's "creativity," because ethics is thought to involve a system of rigid rules of conduct. Others hold that ethics is "only" a matter of right and wrong conduct, so there is not much to teach because scientific misconduct is obviously wrong. Similarly, some think that personal and professional values are essentially synonymous, so there is "nothing special" to learn about scientific research ethics.

However, scientists and engineers encounter a host of ethically laden issues involving research standards and practices that are not usually encountered by lay people. Moreover, however ethically "developed" students may be by the time they enter graduate training, few may have a strong sense of themselves as a moral agent within the professional context of being a scientist. The goal of teaching research ethics is not to "change character" or make "good people." Rather, it is to teach people who want to do good science about the ethical standards and issues in their work, and how to deal with ethical problems that they encounter as scientists. It also is important for students, and their teachers, to learn that scientific misconduct is not a simple matter about which there is little to be said except that it is wrong. As Weil stresses, defining what constitutes fabrication, falsification, and plagiarism and determining whether they have occurred often "turns out to be less straightforward than one might expect."[7] Thus, for example, it is not always easy to distinguish plagiarism from poor citation practices, or to establish that what appears to be fabricated data is the result of an intentional effort to deceive rather than unintentional errors in recording or analyzing findings.

The reasons for and importance of explicit attention to ethical issues in science and engineering have been addressed for many years, well before several highly publicized cases of scientific misconduct catalyzed developments such as the National Institutes of Health (NIH) 1989 requirement that institutions receiving National Service Research Award grants provide trainees with instruction in the responsible conduct of research. In the past decade, major professional organizations have addressed what they have variously called "honor in science," ethical aspects of "being a scientist," and "responsible science."[8] The view that we need to pay ongoing attention to the ethical climate of the scientific research enterprise is supported by the reports of faculty and doctoral students that they are encountering a variety of what they define as questionable or wrong practices in their departments, and the acknowledgment by faculty in those departments that they exercise little collective responsibility for the professional conduct of their colleagues or students.[9]

In February 1994, the Councils of the National Academies of Science and Engineering and the Institute of Medicine affirmed the importance of serious and ongoing attention to research ethics. Their statement called on all participants in the scientific enterprise—"individual researchers, universities and other research institutions, professional organizations, and governments"—"to renew [their] commitment to strengthening the professional climate of the research system." "Research mentors, laboratory directors, department heads, and senior faculty," the Councils stated, "are responsible for explaining, and requiring adherence to high standards."[10]

Beyond the reasons that are particular to the scientific enterprise, the teaching of ethics in higher education serves important and more general goals. Five such goals were discussed in a 1980 Hastings Center report. First, by "stimulating the moral imagination," a course or other ethics teaching format should foster "the ability to gain a feel for the lives of others, some sense of the emotions and the feelings that are provoked by difficult ethical choices, and some insight into how moral viewpoints influence the way individuals live their lives. . ." A second goal is developing "a capacity to sift out [the] ethical issues" embedded in a given situation or problem and "to see the moral implications of individual and collective decisions." Third, "even in the most basic introductory course," participants need to begin developing analytical skills. "Coherence and consistency are minimal goals . . . both in the analysis of ethical propositions and of moral actions and in the justification of rules, principles, and specific moral decisions." Fourth, students "must explore the role, in practice . . . of freedom to make moral choices and personal responsibility" for the choices that are made. The fifth goal is learning to tolerate the disagreements and "inevitable ambiguities in attempting to examine ethical problems." At the same time, "there must be no less an attempt to locate and clarify the sources of disagreement, to resolve ambiguity so far as possible, and to see if ways can be found to overcome differences of moral viewpoint and theory."[11]

IV. HOW TO TEACH RESEARCH ETHICS: SOME SUGGESTIONS

The most feasible and effective methods of teaching research ethics will depend on the educational context or format and the available time. There are many teaching tools for integrating ethics into the content of a disciplinary course, for a format such as "ethics modules" in a course, or for a separate course or seminar series on research ethics. These tools include case studies, minicases or vignettes (including the development by students of new cases), guest speakers, discussions of students' experiences, role playing, games and simulations, films and videos, literature (plays, novels, and short stories), and "cases in the news" that involve ethical problems. One important caveat about the use of audiovisual materials, literature, and discussions that draw on the experience of students or faculty is that, while they can be very effective in meeting the goal of stimulating the moral imagination, they may not include other goals such as developing analytical skills. In the words of the Hastings Center report on teaching ethics, "care must be taken [that such teaching tools] do not swamp the imagination while starving the mind."[12]

A. Case Teaching

Most faculty involved in teaching research ethics have found that case teaching can be an extremely effective method to foster students learning the analytical skills needed to address ethical issues in scientific research, as well as gaining knowledge about the content of those issues. But to be effective, the case method must be taken seriously: It is a demanding task for both instructors and students, involving far more than simply handing out a written case or watching a video and then "talking about it." In a useful short paper, "Hints for Case Teaching," Shapiro notes that "the core of case teaching is the facilitation of student learning," which involves establishing clear objectives, expectations, and responsibilities for both instructors and students.[13]

Interactive group discussions of cases or scenarios, involving faculty, postdoctoral students, graduate students, and research staff, can achieve a number of goals. These goals include (1) transmitting professional values and ethical standards of the discipline to students and in the process clarifying conventions; (2) encouraging the more senior staff to think explicitly about their own standards and values, and their awareness of conventions; and (3) clarifying accepted differences in values and identifying questionable and unacceptable practices.

The value of group discussions about the problematic situations that arise in the course of proposing, conducting, and reporting research is that it helps practitioners at all levels (but especially students and junior researchers) become more aware and knowledgeable about the ethical implications of various aspects of their work. Using a structured format such as the seven-step method outlined later in this chapter, case studies also provide an opportunity to develop reasoning skills, to learn how to analyze situations in a way frequently not possible in the immediacy of real life, to identify the possible range of actions and their potential ramifications, to assess the availability of outside resources, and to draw on the experience, insights, and wisdom of colleagues.

Scenarios or cases that provide a true-to-life example of problematic situations are especially useful as a catalyst to group discussions for a variety of reasons. (1) They demonstrate the reality of situations that raise ethical issues and make the concept less abstract. (2) They serve as a useful vehicle to highlight the fact that everyone has relevant experience to draw on in analyzing the context of the situation, and they facilitate identification and examination of underlying assumptions in a nonthreatening atmosphere. (3) They generally illustrate that because different choices have different consequences for different individuals, which solutions are "better" can depend on one's point of view and stake in how the issue is resolved. (4) Good cases, like real-life situations, have complex interacting features that help to identify sources of confusion and misunderstandings that arise from different assumptions. (5) The ambiguities provide an opportunity to clarify relevant differences that affect decisions regarding the better choice of action and to discuss the range of acceptable practices as well as unacceptable practices. (6) Finally, scenarios can promote an agent-oriented rather than an abstract, judgmental analysis of the situation. As users of the case study method know, often the hardest thing for students to figure out—just as in real life—is how to do the ethically right thing.

Two aspects of cases about which users of case teaching have differing opinions are the pros and cons of actual cases compared to realistic hypothetical cases, and the use of cases or scenarios in a dialogue form compared to a narrative presentation. One strong argument for teaching via a real case, which needs to be carefully formatted with the sources used fully identified, is that it demonstrates that the problems that the case involves actually happen in scientific research, that the issues usually are complex and involve many ambiguities, and that it may be too difficult to gather "the facts" that are important for adequately analyzing and handling the apparent problems. On the other hand, a real case, particularly if it is receiving media attention, carries with it what people may have heard or read about it and what they may know, or think they know, about the alleged facts and the personalities involved.

Turning to the format in which a hypothetical case is presented, some experienced case teachers believe that narrative cases are pedagogically better than dialogued scenarios for learning how to define and analyze ethical issues and develop an ethically justifiable course of action. Proponents of dialogued scenarios, in turn, argue that, as in real life, individuals communicate, or miscommunicate, in the details of normal discourse through oral and body as well as written language. Dialogued scenarios or cases may make it easier for participants to identify differences in underlying assumptions that are frequently the source of confusion and misunderstanding. They can encourage discussions to make explicit different ways of interpreting the same situation, and they make it more difficult for discussants to talk past each other as they may do when they are operating from different, unexpressed assumptions.

Scenarios can be used with any size group. Initially, it is helpful to emphasize for the participants that (1) in discussions of the case there is usually more than one acceptable resolution or course of action; (2) a variety of possible solutions does not mean that every course of action is acceptable (i.e., there is usually at least one unacceptable solution to a problem); (3) the relative advantages or disadvantages of one solution rather than another can vary depending on the individual's perspective (e.g., a graduate student, a postdoctoral fellow, and the lab head may not agree on which course of action is the best); and (4) the participants should consider themselves actors in the scenario, attempting to determine what is the best action to take next, rather than judges of the actions taken by others. In order to decide what to do next, they may need to determine what additional information is needed, what additional resources may be available to draw on, and who else can or should take the next step.

The case should be available to the participants before the discussion, and then summarized at the beginning of the session. With dialogued scenarios, arranging to have different participants play the roles can be a good ice breaker, and in a large group helps the audience keep track of which characters said or did what. It also can make it possible to highlight the nuances of different phrasing.

In large-group formats, presentation of the scenario can be followed by brief (three to five minutes) commentary by each member of a panel drawn from the large group. Each panel member should represent a different constituency, for example, junior faculty, senior faculty, graduate students, postdoctoral trainees, and technical research staff. While a panel member should not be thought of as speaking for all members of their group (e.g., graduate students), each can and should speak from his or her own perspective, briefly identifying concerns raised by the scenario that seem especially significant to them. This process facilitates discussion and reinforces the fact that different individuals may have different, but equally valid, views of the situation.

The discussion leader should identify ahead of time the various issues raised in a given scenario and decide which of those are most important to discuss, since there is invariably more discussion possible than time allows. A large group discussion of a scenario can be followed by informal gathering to encourage further discussion of the issues. Large-group discussion also can be followed by less formal small group discussions (10 to 12 people). The smaller groups encourage those who are uncomfortable speaking before a larger group, or are reluctant to raise an issue in the presence of a superior, an advisor, or an antagonist. Small groups should be composed of representatives of various perspectives—such as students, faculty, and staff, women and men—and should also reflect ethnic and cultural diversity. In addition, there should be attention to separating individuals from the same laboratory or other working group so there is a wide range of experience and perspective represented in the discussion.

Some additional elements to bear in mind for the effective use of case study teaching methods are as follows:

- Define the educational objectives that can be accomplished best by cases, and use other pedagogical methods (e.g., lectures, readings) to meet other goals.
- Select or develop well-constructed cases. For example, cases or mini-cases should (1) be realistic; (2) be concise, since in cases as in real life, we don't know "all the facts" about a situation or problem; (3) involve several types of situations: some that are fairly clear cut, which can help to clarify more complex, murky problems, and some that are suitably ambiguous, presenting quandaries or dilemmas without simple or obvious "right" or "wrong" solutions; (4) involve at least two contrasting ethical positions, which can lead to different resolutions; and (5) not permit a "simple" appeal to laws or regulations to resolve the ethical issues.
- Conduct a structured case analysis, using a method such as the framework described later in this chapter.
- Control but don't dominate the case session; the analysis and discussion should be driven by the participants, especially students.
- Try to have the participants reach a decision on the case, as they will have to do when dealing with ethical issues in real life.
- Have participants give arguments that defend the action plans they develop, or courses of action that they believe are ethically unacceptable.
- Don't pretend to be value-neutral; facilitators can also take positions on the issues and recommend a course of action.[14]

B. Teaching Research Ethics In Situ

Members of laboratory and other research groups have informal and usually impromptu discussions about ethical issues that arise in the course of their work, such as authorship policies or how data should be handled, as well as discussions about cases of alleged scientific misconduct that are in the news or known to members of the groups. These conversations may be led by the head of the research group, or may take place more privately among graduate students and/or postdocs, in or out of the research setting.

Such discussions give participants a chance to air their concerns about perspectives, but generally do not provide a systematic framework for learning about research ethics. Building on these types of interactions, we believe that a particularly effective way to teach and learn research ethics would be to incorporate structured discussions about frequently encountered ethical issues into the everyday life of the laboratory or other research training sites, such as field work in engineering. Creating time for members of a research group to meet regularly to examine such issues in relation to the group's work and their larger disciplinary community is a "real-life" version of the case study method. Additionally, bringing to the surface and discussing matters that often simmer beneath the surface in a lab can prevent the eruption of "unanticipated" problems in the conduct and work of individuals and in interpersonal relationships within a research group.

To our knowledge, this format and method have not been used often for scientific research ethics. But we think it is an effective instructional method, particularly if the directors of research groups have a faculty development experience and if students have had some prior curriculum exposure to applied ethics. In this context, as well as in the classroom, team teaching is a possible strategy. One such team-teaching approach would be to have faculty in the humanities or philosophy do "ethics rounds" with lab groups, modeled on the bedside clinical ethics teaching utilized by many medical schools and teaching hospitals for teaching medical ethics to students and residents.

C. A Framework for Reasoning About Ethical Issues in Research

One of many possible frameworks for teaching and learning applied ethics that has been well tested in the classroom has been developed for business ethics.[15] This three-part framework provides a basic approach to teaching ethical reasoning skills, particularly when used in conjunction with a case study method. In this section, we summarize the framework in outline form, and illustrate how its seven-step procedure can be used with a case-study teaching format. The framework focuses on learning how to use ethical principles and moral reasoning skills to identify, analyze, and decide how best to respond to ethical issues. Drawing from the Andersen manual and adaptations of those materials provided by Patricia Werhane and Vivien Weil, the basic elements of the "three-part framework for moral reasoning" are as follows:

1. Identify and sort out the ethical issues or questions in the situation, and the levels of analysis they involve.

- What is the problem(s)?
- What makes it an ethical issue? Is there more than one ethical issue?
- Who is affected or involved, and what levels of analysis are needed? For example, what ethical issues are raised for the individual, the professional group, the institution or organization, and the larger social system in which these entities function?
- Are there different ways to describe or characterize the ethical problems(s)?
- What values or standards are at stake?

2. Decide on the ethical principles or standards that will be applied to the ethical analysis and decision making.

While there are several major philosophical approaches that faculty and students can draw on in examining a given ethical problem, we do not believe that those engaged in teaching and learning research ethics must become experts in various ethical theories or sets of principles. Rather, recognizing that ethical problems, and their examination, are complex, we advocate a "commonsense morality" approach. This approach involves thinking about the ethical standards or rules of conduct that you ideally would like someone to use in order to achieve what you view as the ideal or most desirable outcome or resolution of the issue under consideration. In this reasoning process, find out whether there is a relevant professional code of ethics, and if so, whether you agree with its standards and find them useful in your analysis.

3. Follow a seven-step procedure for ethical decision making.

As Weil[16] emphasizes, this is a structured procedure [developed by Velasquez[17] for business ethics] which provides "a kind of checklist,

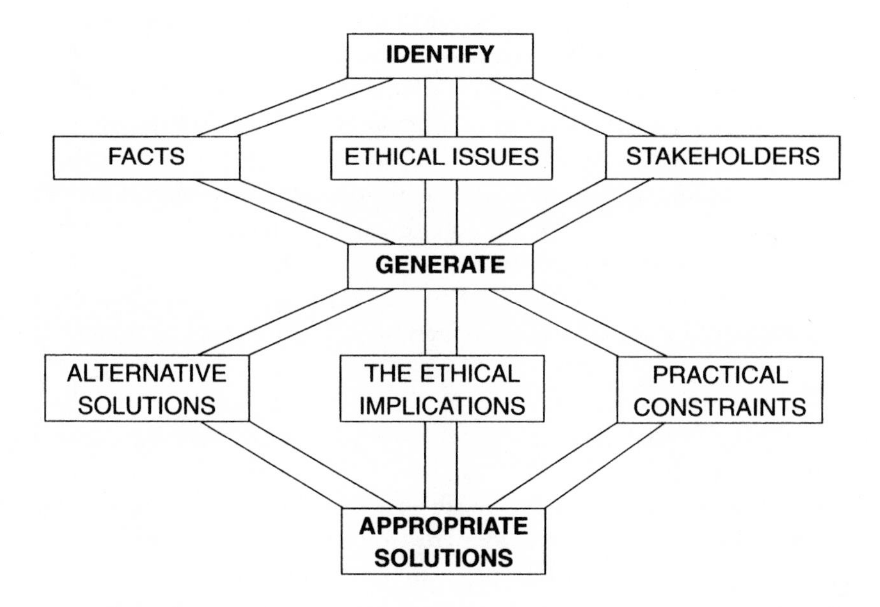

not a mechanical decision procedure," and pushes people to reach closure on how they will handle the problem.

1. Recognize and define the ethical issues (see 1).
2. Identify the key facts involved in the situation, and also the factual ambiguities or uncertainties and what essential additional information is needed.
3. Identify the affected parties or "stakeholders." The stakeholders can be individuals and groups, who affect and are affected by the way the ethical problem is resolved. In a case involving intentional deception in reporting research findings, for example, the stakeholders can include the individual or individuals who perpetrated the deception, other members of their research group, their department and university, the funder, the journal that has published the results, and other researchers developing or conducting research based on the findings.
4. Formulate viable alternative courses of action that could be taken, and continue to check the facts.
5. Assess the alternatives. For example:
   - What are the ethical implications of each alternative: Is it in accord with the ethical standards being used? If not, can it be justified, and on what grounds? What consequences will it have for the affected parties? What issues will it leave unresolved? Can the

course of action be publicly defended on ethical grounds? What kind of precedent would the alternative set?
- What practical constraints are involved in adopting each alternative? Constraints may include lack of knowledge or uncertainty, lack of ability or authority or resources to implement the course of action, or institutional, structural, or procedural barriers.

6. Construct desired options and persuade or negotiate with others to implement the options.
7. Decide what actions should be taken to implement the chosen course of action. In so doing, recheck and weigh your reasoning in steps 1 through 6.

The next section illustrates how the seven-step method can be applied to case teaching, using, in this instance, a dialogued scenario. In working through steps 4 through 7, with this scenario or any case, we recommend that the participants adopt the role of a moral agent rather than that of a judge. That is, participants should be encouraged to consider and discuss what they would do as a principal in the case, rather that what the principals should have done.

4. Busman's Holiday: Applying the Seven-Step Procedure to a Case Study Discussion

BUSMAN'S HOLIDAY[18]

Participants: John Cooper—A graduate student in neuropharmacology at MIT
Bill Steep—A graduate student in molecular biology at Cal Tech

[*Three years after graduating from Williams College, where they were close friends and roommates, John and Bill meet during Christmas vacation at John's mother's house in Newton, Massachusetts.*]

JOHN: I couldn't believe it when Susan told me you were in town visiting your sister. It's been so long since we've had a chance to talk.

BILL: I guess neither of us has much time to write. I'd forgotten that Susan went to your high school. I call her occasionally to check up on the old Williams crowd. She told me that you really like MIT.

JOHN: It's great. My advisor, Alan Schneider, is top notch. The lab works on neurohormones. I'm trying to develop a strategy to enhance the secretion of endorphins in postsurgical patients.

BILL: That's interesting. I actually know a little bit about that because there's a group on my floor that's looking at secretory pathways in the pituitary.

In fact, one of my best friends in the department, Lynne Thomas, works in that group. We started at Cal Tech at the same time. Lynne works closely with a postdoc named Jim Taylor. Have you met him?

JOHN: I've heard of him, but we've never met. I think his graduate thesis dealt with endorphin antagonists.

BILL: That's right. And now he and Lynne are working on two different aspects of endorphin synthesis and secretion. Jim thinks he's found a new enzyme that helps regulate the availability of endorphins for secretion. I'll know more about it next month—he's scheduled to give the January seminar.

JOHN: That's fascinating. I should give him a call. Just knowing that the enzyme exists might help us solve a puzzle we've encountered in the data. We've found a cofactor that enhances endorphin production, but we don't know how it works.

BILL: Lynne is working on a possible cofactor too. I'm sure she'd be very interested in your work.

JOHN: Tell her to give me a call. Our first paper has just been accepted by *Cell.* By the way, do you know anything about the structure of Jim's new enzyme?

BILL: Yes, I've heard it has a very unusual tertiary structure with . . .

[*Bill stops abruptly and appears to look at something over John's shoulder*]

JOHN: With what? What's wrong?

BILL: You know, it's great to see you. I'd forgotten how much I enjoy talking science with you. But I just realized that Jim Taylor is very cautious about sharing his data; in fact, I've never seen anyone so sensitive about being scooped. I don't know how he'd feel about us having this conversation. I don't think he'd like it—especially since he's beginning to look for a permanent position.

JOHN: Sorry. I didn't mean to make you uncomfortable. Um . . . did you see the Celtics game last night? I guess they're out of the playoffs again.

BILL: I guess so. It's depressing isn't it? Two weeks ago, Lynne and I got tickets for a Lakers game. Do you know that Vlade Divac . . .

[*The discussion continues but doesn't return to science.*]

We assume that everyone participating in the case discussion will have studied the case and reviewed the framework before the case is analyzed in a group setting. The case should be presented in its dialogued form, or recapped by the discussion leader/facilitator. The group then examines the case, with the facilitator using steps such as the following to guide the discussion and help the participants reach closure on how they would handle the issues they identify. We have included some examples of the types of points that the discussion leader should try to elicit from the participants.

Step 1. The participants should begin by identifying the issues or problems in the case, and explaining why they constitute ethical problems. A central issue, for example, is that of rights and responsibilities concerning intellectual property with regard to the sharing of work-in-progress techniques or data, in an ethical rather than legal sense. To what extent is Jim's new enzyme, and knowledge of its structure, "his property"—before he makes a presentation or report about it? Given the value of communality that is supposed to be part of the scientific ethos, does Bill have a "right" to share any of this information with his friend, John, knowing that John's lab group is working in the same research area? Similar questions can be asked about the cofactors that John's and Lynne's groups are working on.

Step 2. Next, have the participants identify the key facts involved in this situation: for example, information about what research topics are being pursued by which lab groups, and which lines of research, discoveries, or new data by each group would be of interest to another group. As part of this step, participants also should identify what is not known about the elements in the case, and what sorts of assumptions they are making due to these unknowns.

This part of the analysis helps participants learn that in hypothetical cases, as in real life research, ambiguities and presuppositions—such as what the "facts" are—can cause a high level of disagreement about how the issues should be dealt with. How a case is addressed and dealt with, participants learn, will vary depending on how they fill in the unknowns.

Step 3. Have the participants identify the affected parties or "stakeholders" in this case, both individuals and groups, and specify what they have at stake. For example, what does John, as a graduate student, stand to gain or lose if he decides that he should tell, or not tell, his advisor what he has learned from Bill about the Cal Tech group's work? Similarly, will Bill's work and professional and personal relationships at Cal Tech be affected if he tells Lynne or postdoc Jim Taylor about his conversation with John? What may be at stake for Jim Taylor, as a postdoc, if the MIT lab learns about his new enzyme?

Step 4. Based on the participants' definition of the ethical problems in the case, the key facts involved, and their specification of the affected parties and what they have at stake, the group next should formulate some alternative courses of action that could be taken by John and Bill, and, depending on these courses of action, by other involved parties.

Some examples of courses of action might be:

- John recontacts Bill and they agree that their conversation about the work of the MIT and Cal Tech lab groups should not be discussed with anyone else.
- John tells his advisor what he has learned about the Cal Tech group's work. His advisor then . . . [lay out alternative actions Dr. Schneider might take].
- Bill discusses the work of John's MIT group with (a) Lynne and/or (b) Jim when he returns to Cal Tech. Lynne or Jim then . . . [layout alternative courses of action Lynne or Jim might take].

Step 5. After what seem to be some viable alternatives are developed (not a lengthy list of every conceivable and sometimes very unlikely course of action), each alternative needs to be assessed both ethically and practically. For example, what are the ethical implications of each of the illustrative examples noted in step 4? Are these implications in accord with the ethical standards that participants believe should be applied to this case? Is there consensus among members of the group that the standards and preferred course of action are ethically acceptable according to "commonsense" morality? What consequences will adopting a given course of action have for the stakeholders? What issues will it leave unresolved? Would the participants be willing and able to publicly defend the course of action on ethical grounds? What kind of precedent would the alternative set?

As is true when dealing with real-life research ethics problems, the participants also need to be aware of the kinds of practical constraints that are involved in adopting possible courses of action. Constraints that need to be identified in dealing with a particular case may include lack of knowledge or uncertainty, lack of ability or authority or resources to implement a course of action, or institutional, structural, or procedural barriers.

Step 6. Having assessed the alternatives, the participants next should try to reach agreement, or perhaps a majority and minority opinion, on the most desirable or best course of action. (In some particularly difficult cases, it can be a question of which is the least bad, not the best, course of action.) This step probably will involve negotiating with others in the group over both ethically and practically based disagreements about how the problem should be handled.

Step 7. Finally, having agreed on a course of action, the group needs to decide what actions should be taken to implement it and thus reach the desired outcome. In this process, it is helpful to recheck with the reasoning that the participants used in steps 1 through 6.

Notes

1. Patricia Werhane and Vivian Weil made important contributions to this chapter, particularly through providing us materials from their courses.

2. J. P. Swazey, "Teaching ethics: needs, opportunities, and obstacles," in *Ethics, Values, and the Promise of Science. Sigma Xi Forum Proceedings.* February 25–26 (Research Triangle Park, N.C.: Sigma Xi, 1993), 238.

3. J. R. Rest, "Can ethics be taught in professional schools? The psychological research," *Easier Said than Done* (1988); See also M. J. Bebeau, "Can ethics be taught? A look at the evidence," *Journal of the American College of Dentists* 58(1) (1991); T. R. Piper, M. C. Gentile, and S. D. Parks, *Can Ethics Be Taught?* (Boston: Harvard Business School, 1993); J. R. Rest, "Moral development in young adults," in *Adult Cognitive Development: Methods and Models,* ed. R. A. Mines and K. S. Kitchener (New York: Praeger, 1986) 92–111; M. Trow, "Higher Education and Moral Development," *AAUP Bulletin* 61(4) (1976).

4. J. P. Swazey, K. S. Louis, and M. S. Anderson, "University policies and ethical issues in research and graduate education: Highlights of the CGS Dean's Survey," *CGS Communicator* 23(3) (1989); Swazey, "Teaching ethics."

5. B. Lo, "Skepticism about teaching ethics," in *Ethics, Values, and the Promise of Science,* 151–53.

6. Swazey, "Teaching ethics." 227.

7. V. Weil, "Teaching ethics in science," in *Ethics, Values, and the Promise of Science,* 224.

8. Institute of Medicine Committee on the Responsible Conduct of Research, *The Responsible Conduct of Research in the Health Sciences* (Washington, D.C.: National Academy, 1989); Panel on Scientific Responsibility and the Conduct of Research, *Responsible Science: Ensuring the Integrity of the Research Process,* Vols. 1 and 2 (Washington, D.C.: National Academy Press, 1992); I. Jackson, *Honor in Science* (New Havaen, Conn.: Sigma Xi, 1986).

9. J. P. Swazey, M. S. Anderson, and K. S. Louis, "Ethical problems in academic research," *American Scientist* 81 (1993).

10. Councils of the National Academy of Sciences, National Academy of Engineering, and Institute of Medicine, "Statement on scientific conduct," 1994.

11. Hastings Center, *The Teaching of Ethics in Higher Education* (Hastings-on-Hudson, N.Y.: Hastings Center, 1980), 48–52.

12. Hastings Center, *The Teaching of Ethics,* 1980.

13. B. P. Shapiro, *Hints for Case Teaching* (Boston: Harvard Business School, 1984).

14. For more detailed suggestions about effective case teaching, see A. Anderson and S. C. Co, "Business Ethics Program" [Ethics foundation presentation and overheads; Classroom integration: concepts and techniques.] (1992); Lo, "Skepticism"; S. J. Bird, "Teaching ethics in science: Why, how, and what," in *Ethics, Values, and the Promise of Science,* 228–32; Shapiro, "Hints."

15. Andersen and Co, "Business Ethics."

16. Weil, "Teaching ethics."

17. M. Velasquez, *Business Ethics: Concepts and Cases,* third edition (Englewood Cliffs, N.J.: Prentice Hall, 1992).

18. Scenario by Eve K. Nichols and Stephanie J. Bird.

2

Morality and Scientific Research[1]

BERNARD GERT

Any useful attempt to resolve the moral problems that may arise in the course of performing, proposing, or publishing scientific research requires a clear and explicit account of morality. This chapter attempts to provide such an account of morality. It is not an attempt to revise our standard or common morality, simply an attempt to describe it. Common morality does not provide a unique solution to every moral problem, but it always provides at least a way of distinguishing between morally acceptable answers and those that are morally unacceptable—that is, it places a significant limit on legitimate moral disagreement.

I. AREAS OF MORAL AGREEMENT

One reason for the widely held belief that there is no common or standard morality is that the amount of disagreement in moral judgments is vastly exaggerated. However, everyone agrees that such kinds of actions as killing, causing pain or disability, and depriving of freedom or pleasure are immoral unless one has an adequate justification for doing these kinds of actions. Similarly, everyone agrees that deceiving, breaking a promise, cheating, breaking the law, and neglecting one's duties also need justification in order not to be immoral. No one has any real doubts about this. There is sometimes disagreement about whether a particular act counts as deceiving, but once one has determined that an act counts as deceptive, there is agreement that it needs to be justified. Although there is also some disagreement on what counts as an adequate moral justification for any particular act of deceiving, there is overwhelming agreement on some features of an adequate justification. Everyone agrees that what counts as an adequate justification for one person must be an adequate justification for anyone else when all of the morally relevant features of the situation are the same. This is part of what is meant by saying that morality requires impartiality.

There is also agreement that it is not the case that only the most sophisticated philosopher can understand what counts as an adequate justification for deception. No one engages in a moral discussion of questions like "Is it morally acceptable to falsify data in order to increase one's chances of getting a large grant?" because everyone knows that such deception is not justified. Morality is learned by children starting at a very early age, and by the time they are in high school most children know in most cases whether acting in a certain way is morally acceptable or not. This is part of what is meant by saying that morality is a public system.

Finally, everyone agrees that the world would be a better place if everyone acted morally, and that it gets worse as more people act immorally more often. This is why one should try to teach everyone to act morally even though this effort will not be completely successful. Although in particular cases a person might benefit personally from acting immorally, e.g., falsifying data in order to get a grant when there is almost no chance of being found out, even in these cases it is not irrational to act morally, viz., to present the correct data even though it means one will most likely not get the grant. Morality is not determined by feelings or by mystical revelation, it is the kind of public system that every rational person supports. This is part of what is meant by saying that morality is rational.

II. RATIONALTY AS AVOIDING HARMS AND NOT AVOIDING BENEFITS

In this section I try to provide an account of rationality that explains its relationship to morality and self-interest. Everyone agrees that unless one has an adequate reason for doing so, it would be irrational not to avoid any harm or to avoid any benefit. The present account of rationality, although it accurately describes the way in which the concept of rationality is ordinarily used, differs radically from the accounts normally presented in two important ways. First, it starts with irrationality rather than rationality, and second, it defines irrationality by means of a list rather than a formula. The basic definition is as follows: *A person acts irrationally when that person acts in a way that he or she knows (justifiably believes), or should know, will significantly increase the probability that he or she will suffer any of the items on the following list: death, pain, disability, loss of freedom, or loss of pleasure; and he or she does not have an adequate reason for so acting.*

The close relationship between irrationality and harm is made explicit by this definition, for this list also defines what counts as a harm or an evil. Everything that anyone counts as a harm or an evil, such as disease or punishment, is related to one or more of the items on this list. These items are broad categories, so that nothing is ruled out as a harm or evil that is normally regarded as such. However, except for death, all of these harms have degrees, and even the time of one's death can vary greatly. Further, although there is complete agreement on all of the basic harms, there is no universal agreement on the ranking of these harms. Complete agreement on what the basic harms or evils are is compatible with considerable disagreement on what is the lesser of two evils.

Since having an adequate reason can make harming oneself rational, a full understanding of rationality requires understanding not only what counts as a reason, but also what makes a reason adequate. The basic definition is as follows: *A reason is a conscious belief that one's action will help anyone, not merely oneself or those one cares about, avoid one of these harms, or gain some good, namely, ability, freedom, or pleasure, and such that this belief is not seen to be inconsistent with one's other beliefs by almost everyone with similar knowledge and intelligence.* What was said earlier about evils or harms also holds for the goods or benefits mentioned in this definition of a reason. Everything that anyone counts as a benefit or a good, such as health, love, or friends, is related to one or more of the items on this list or to the absence of one or more of the items on the list of evils. Complete agreement on what counts as a good is compatible with considerable disagreement on whether one good is better than another, or whether gaining a given good or benefit adequately compensates for suffering a given harm or evil.

A reason is adequate if any significant group of otherwise rational people regard the harm avoided or benefit gained as at least as important as the harm suffered. People are otherwise rational if they do not knowingly suffer any harm without some reason. Just as no beliefs held by any significant religious, national, or cultural group are regarded as delusions or irrational beliefs—for example, the belief held by Jehovah's Witnesses that accepting blood transfusions will have bad consequences for one's afterlife is not regarded as an irrational belief or delusion—so no rankings that are held by any significant religious, national, or cultural group are regarded as irrational—for example, ranking the harms that would be suffered in an afterlife as worse than dying decades earlier than one would have if one accepted a transfusion is not an irrational ranking. The intent is to not rule out as an adequate reason any relevant belief that has any plausibility; the goal is to provide an account of irrationality on which there is close to universal agreement that no one ever wants anyone he or she cares for to act irrationally. Only such an account makes irrationality an objective concept and prevents the term *irrational* from degenerating into a term of general disparagement.

This account of rationality, although it may sound obvious, is in conflict with the most common account of rationality, where rationality is limited to an instrumental role. A rational action is often defined as one that maximizes the satisfaction of all of one's desires, but without putting any limit on the content of those desires. This results in an irrational action being defined as any action that is inconsistent with such maximization. But unless desires for any of the harms on the list are ruled out, it turns out that people would not always want those for whom they are concerned to act rationally; for example, no one wants a loved one who is suffering from a mental disorder to maximize the satisfaction of his or her desires if this involves self-mutilation and suicide.[2] That rationality has a definite content and is not limited to an instrumental role, such as acting so as to maximize the satisfaction of all one's desires, goes contrary to most accounts of rational actions.[3]

Scientists may claim that both of these accounts of rationality are misconceived. They may claim that the basic account of rationality does not regard rationality as primarily related to actions at all, but rather regards rationality as reasoning correctly. Scientific rationality consists of using those scientific methods best suited for discovering truth. Although I do not object to this account of rationality, I think that it cannot be taken as the fundamental sense of rationality. The account of rationality as avoiding harms is more basic than that of reasoning correctly, or scientific rationality. Scientific rationality cannot explain why it is rational to avoid suffering avoidable harms when no one benefits in any way. The avoiding harm account of rationality does explain why it is rational to reason correctly and to discover new truth, namely, because doing so helps people to avoid harms and to gain benefits.

It is very important for scientists to realize that the fundamental sense of rationality involves the avoidance of harms, not the seeking of new truth. Not that the seeking of truth is unimportant—we could not avoid harm if we did not know the truth about the world. But to take truth to be the ultimate goal, regardless of the harms involved in obtaining it, is as irrational as the miser taking money to be the ultimate goal. Truth is primarily an instrumental value, although for many of the best scientists it often takes on an aesthetic value as well; they take pleasure in discovering truth for its own sake. However, neither rationally nor morally does seeking truth confer any special status on an action. Doing scientific research is governed by the same general rational and moral constraints as any other kind of activity.

On the account of rationality I have presented, when there is a conflict between morality and self-interest, it is not irrational to act in either way. Although this means that it is never irrational to act contrary to one's own best interests in order to act morally, it also means that it is never irrational to act in one's own best interest even though this is immoral. Further, it is not irrational to act contrary to both self-interest and morality, for example, if friends, family, or colleagues benefit. This latter fact is often not realized, and some physicians and scientists feel that if they act to benefit others and contrary to their own self-interest, they cannot be acting immorally. This allows them to immorally cover up the mistakes of their colleagues, believing that they are acting morally, because they themselves have nothing to gain and are even putting themselves at risk. Indeed, the misunderstanding of the relationship of loyalty to morality is one of the most important practical issues in moral philosophy.

Although some philosophers have tried to show that it is irrational to act immorally, this conflicts with the ordinary understanding of the matter. Everyone agrees that in some circumstances it may be rational for someone to deceive in order to get a grant, even if this is acting immorally. Nowhere in this chapter do I attempt to provide the motivation for a person to act morally. That motivation primarily comes from one's concern for others, together with a realization that it would be arrogant to think that morality does not apply to oneself and one's colleagues in the way that it applies to everyone else. The attempt to provide a useful guide for determining what ways of behaving are morally acceptable when one is confronted with a moral problem presupposes that most of the scientists who read this chapter want to act morally.

III. MORALITY AS A PUBLIC SYSTEM

A public system is a system that has the following characteristics. (1) All persons to whom it applies, those whose behavior is to be guided and judged by that system, understand it, that is, know what behavior the system prohibits, requires, encourages, and allows. (2) It is not irrational for any of these persons to accept being guided or judged by that system. The clearest example of a public system is a game. A game has an inherent goal and a set of rules that form a system that is understood by all of the players—that is, they all know what kind of behavior is prohibited, required, encouraged, and allowed by the game, and it is not irrational for all players to use the goal and the rules of the game to guide their own behavior and to judge the behavior of other players by them. Although a game is a public system, it applies only to those playing the game. Morality is a public system that applies to all moral agents; all people are subject to morality simply by virtue of being rational persons who are responsible for their actions.

In order for morality to be known by all rational persons, it cannot be based on any factual beliefs that are not shared by all rational persons. Those beliefs that are held by all rational persons include general factual beliefs such as: people are mortal, can suffer pain, can be disabled, and can be deprived of freedom or pleasure; also, people have limited knowledge, that is, people know some things about the world, but no one knows everything. On the other hand, not all rational people share the same scientific and religious beliefs, so that no scientific or religious beliefs can form part of the basis of morality itself, although, of course, such beliefs are often relevant to making particular moral judgments. In a parallel fashion, only personal beliefs that all rational persons have about themselves, such as beliefs that one can be killed and suffer pain, can be used in providing a foundation for morality. Excluded as part of a foundation for morality are all personal beliefs about one's race, sex, religion, and so on, because these beliefs are not shared by all rational persons.

Although morality itself can be based only on those factual beliefs that are shared by all rational persons, individual moral decisions and judgments obviously depend not only on the moral system but also on the situation. In fact, most actual moral disagreements, such as whether a particular scientist acted properly or not, are based on a disagreement on the facts of the case, for example, whether or not the scientist knowingly used information gained from reviewing a grant, to formulate his or her own grant proposal. Other moral disagreements depend upon disagreements on what standards are appropriate to apply—for example, should a competent scientist have known that the experiment posed a significant risk of harm? Still others may depend upon disagreements about what counts as breaking a rule, such as when not reporting failed experiments counts as deception. Thus, in order to be qualified to make a moral judgment in a particular field, one must know the conventions and practices of that field.

Almost all the difficult moral cases that arise in the area of scientific research depend upon determining whether the action or practice under consideration is one that needs justification, for example, is deceptive. Unlike the field of medicine, where determining that the action needs justification is often just the start of the moral inquiry (e.g., one has to decide whether the benefit to the patient justifies deceiving the patient), in scientific research it is only rarely that people attempt to justify what they acknowledge to be deceptive (e.g., deception experiments in psychology). Generally, the issue turns completely on whether or not the behavior is properly characterized as deceptive, such as using a nonstandard statistical method that produces more significant results or listing as an author someone who has not even seen the paper. Thus, for most moral issues that arise in scientific research, there is no need for a sophisticated philosophical account of morality. In science, everyone acknowledges that deception is unjustified and the whole discussion turns on whether or not the action or practice counts as deceptive. That is why knowledge of the field is so important, for most of the reasoning involved concerns whether or not people are misled or deceived by the action or practice. Here, knowledge of what scientists and consumers of science believe is more useful than knowledge of morality. The framework of our common morality can be applied to a scientific practice only after it is clear how a particular act or practice should be characterized. However, knowledge of the moral framework is helpful in showing that there is no special morality for scientists.

Although morality is a system that is known by all those who are held responsible for their actions, it is not a simple system. A useful analogy is the grammatical system used by all competent speakers of a language. Almost no competent speaker can explicitly describe this system, yet they all know it in the sense that they use it when speaking themselves and in interpreting the speech of others. If presented with an explicit account of the grammatical system, competent speakers have the final word on its accuracy. They should not accept any description of the grammatical system if it rules out speaking in a way that they regard as acceptable or allows speaking in a way that they regard as completely unacceptable.

In a similar fashion, a description of morality or the moral system that conflicts with one's own considered moral judgments normally should not be accepted. However, an explicit account of the systematic character of morality may make apparent some inconsistencies in one's moral judgments. Moral problems cannot be adequately discussed as if they were isolated problems whose solution did not have implications for all other moral problems. Providing an explicit account of morality may reveal that some of one's moral judgments in one area are inconsistent with the vast majority of one's other judgments. Thus, one may come to see that what was accepted by oneself as a correct moral judgment is in fact mistaken. Even without challenging the main body of accepted moral judgments, particular moral judgments, even of competent people, may sometimes be shown to be mistaken, especially when long-accepted ways of thinking are being challenged. In these situations, one may come to see that one was misled by superficial similarities and differences and so was led into making judgments that are inconsistent with the vast majority of one's other moral judgments. For example, many scientists have recently discovered that their moral judgments about what was morally allowable regarding who should be listed as an author of an article are inconsistent with the vast majority of their other moral judgments.

As noted earlier, there are certain kinds of actions that everyone regards as being immoral unless one has an adequate justification for doing them. These kinds of actions are killing, causing pain or disability, depriving of freedom or pleasure, deceiving, breaking a promise, cheating, breaking the law, and neglecting one's duty. Anyone who kills people, causes them pain, deceives them, or breaks a promise, and does so without an adequate justification, is universally regarded as acting immorally. Saying that there is a moral rule prohibiting a kind of act is simply another way of saying that a certain kind of act is immoral unless it is justified. Saying that breaking a moral rule is justified in a particular situation, such as breaking a promise in order to save a life, is another way of saying that a kind of act that would be immoral if not justified, is justified in this kind of situation. When no moral rule is being violated, saying that someone is following a moral ideal, such as relieving pain, is another way of saying that the person is doing a kind of action regarded as morally good. Using this terminology allows one to formulate a precise account of morality, showing how its various component parts are related.[4] Such an account may be helpful to those who must confront the problems raised by the practices that they encounter in performing, proposing, or publishing scientific research.

IV. JUSTIFYING VIOLATIONS OF THE MORAL RULES

Almost everyone agrees that the moral rules are not absolute, that all of them have justified exceptions; most agree that even killing is justified in self-defense. Further, one finds almost complete agreement on several features that all justified exceptions have. The first of these involves impartiality: Everyone agrees that all justified violations of the rules are such that if they are justified for any person, they are justified for every person when all of the morally relevant features are the same. The major, and probably only, value of simple slogans like the Golden Rule, "Do unto others as you would have them do unto you," and Kant's Categorical Imperative, "Act only on that maxim that you would will to be a universal law," is as heuristic devices designed to get one to act impartially when one is contemplating violating a moral rule. In trying to decide what to do in difficult cases, however, it is more useful and less likely to be misleading to consider whether one would be prepared to publicly allow that kind of violation, that is, allow it to be included in that public system that is morality.

Acting in an impartial manner with regard to the moral rules is analogous to a referee impartially officiating a basketball game, except that the referee is not part of the group toward which he or she is supposed to be impartial. The referee judges all participants impartially if he or she makes the same decision regardless of which player or team is benefited or harmed by that decision. All impartial referees need not prefer the same style of basketball; one referee might prefer a game with less bodily contact, hence calling more fouls, while another may prefer a more physical game, hence calling fewer fouls. Impartiality allows these differences as long as the referee does not tell only one team of these preferences and does not favor any particular team or player over any other in calling fouls. In the same way, moral impartiality allows for differences in the ranking of various harms and benefits as long as one would be willing to make these rankings public and one does not favor any one rational person or group of persons, including oneself and one's friends, over any others when one decides whether to violate a moral rule or judges whether a violation is justified.

The next feature on which there is almost compete agreement is that it has to be rational to favor everyone being allowed to violate the rule in these circumstances. Suppose that some person suffering from a mental disorder both wants to inflict pain on others and wants pain inflicted on him- or herself. The person favors allowing all persons to cause pain to others if they would not complain if others caused pain to them. This is not sufficient to justify that kind of violation. No impartial rational person would favor allowing those who don't complain when they are caused pain to cause pain to everyone else. The result of allowing that kind of violation would be an increase in the amount of pain suffered with no benefit to anyone, which is clearly irrational.

Finally, there is general agreement that the violation should be publicly allowed. To justify a violation it is not sufficient that it would be rational to favor allowing everyone to violate the rule in the same circumstances, if one favors it only if almost no one knows that it is allowable to violate the rule in those circumstances. For example, when almost no one knows that such deception is allowed, it might be rational for one to favor allowing a great scientist to deceive others by claiming to have greater confirmation than he or she actually has if failing to make this false claim is likely to lead other scientists to give up on a theory that the scientist is absolutely convinced is true. But that would not make deception in these circumstances justified. It has to be rational to favor allowing this kind of deception when everyone knows that it is allowed to deceive in these circumstances. The requirement that the violation be publicly allowed guarantees the kind of impartially required by morality.

Not everyone agrees on which violations satisfy the three conditions of impartiality, rationality, and publicity, but it is part of our moral system that no violation is justified unless it satisfies all three of these conditions. Acknowledging the significant agreement concerning justified violations of the moral rules, while allowing for some disagreement, results in the following formulation of the appropriate moral attitude toward violations of the moral rules: *Everyone is always to obey the rule unless an impartial rational person can advocate that violating it be publicly allowed. Anyone who violates the rule when no impartial rational person can advocate that such a violation be publicly allowed may be punished.* (The "unless clause" only means that when an impartial rational person can advocate that such a violation be publicly allowed, impartial rational persons may disagree on whether or not one should obey the rule. It does not mean that they agree one should not obey the rule.)

Anyone acting or judging as an impartial rational person decides whether or not to advocate that a violation be publicly allowed by estimating what effect this kind of violation, if publicly allowed, would have. However, rational persons, even if equally informed, may disagree in their estimate of whether more or less harm will result from this kind of violation being publicly allowed. Disagreements in the estimates of whether a given kind of violation being publicly allowed will result in more or less harm may stem from two distinct sources. The first is a difference in the rankings of the various kinds of harms. If someone ranks a specified amount of pain and suffering as worse than a specified amount of loss of freedom, and someone else ranks them in the opposite way, then although they agree that a given action is the same kind of violation, they may disagree on whether or not to advocate that this kind of violation be publicly allowed. The second is a difference in estimates of how much harm would result from publicly allowing a given kind of violation, even when there seems to be no difference in the rankings of the different kinds of harms. These differences may stem from differences in beliefs about human nature or about the nature of human societies. Insofar as these differences cannot be settled by any universally agreed upon empirical method, such differences are best regarded as ideological. But sometimes there seems to be an unresolvable difference when a careful examination of the issue shows that there is actually a correct answer.

Notes

1. Reprinted, with permission from Sigma Xi, from *Ethics, Values, and the Promise of Science. Sigma Xi Forum Proceedings.* February 25–26 (Research Triangle Park, N.C.: Sima Xi, 1993), 157–65. The original article in Sigma Xi contains a discussion of issues raised by David Goodstein in his very interesting and provocative article, "Scientific fraud," *American Scholar* 60(4) (1991):505–15.

2. See "Irrationality and the DSM-III-R Definition of Mental Disorder," *Analyse & Kritik* 12(1) (1990):34–46.

3. See "Rationality, human nature, and lists," *Ethics* 100(2, Jan. 1990):279–300.

4. For a fuller account, see Bernard Gert, *Morality: A New Justification of the Moral Rules* (New York: Oxford University Press, 1988).

3

Examples of Scientific Misconduct

ALLAN U. MUNCK

Scientific misconduct comes in endless varieties that violate virtually every moral precept. This chapter presents a few well-known modern cases that illustrate not only some of the forms that scientific misconduct has taken, but also the central role, for better or worse, of individual and institutional responses to such incidents.

At the heart of most cases of scientific misconduct is a lone scientist—a student or postdoctoral fellow, a young independent researcher, a well-established senior investigator—along with a set of experiments or observations. What can motivate such a person, steeped in a culture that above all else values truth, to violate the truth by knowingly reporting those experiments or observations falsely? "Truth," of course, is a high-flown word not often used by scientists. In more prosaic terms it might simply mean the actual result yielded by an experiment, as distinct, for example, from the result the experimenter had hoped for. It is the coin of the scientific realm, passed from one scientist to another and added to the repository of reliable facts on which new theories, hypotheses and experiments are based. On the one hand "truth," as applied to scientific research, is a romantic ideal; on the other, it is a down-to-earth necessity without which scientific progress and the practical applications of science would be impossible. Its value can be debased in many ways: by sloppy procedures, by careless description of experimental methods, by biased account of results, but most perniciously by deliberate misrepresentation or fraud.

Why misrepresent experiments or observations? Aside from the murky psychological undercurrents that spawn extremes of misconduct like those referred to later as "professional" fraud, the temptations and pressures that beset most scientists and overwhelm a few of them are not hard to identify. To start with, there are the simple emotions that are an integral part of doing research at every level—the joy of discovering something new, the pleasure of demonstrating one's own ideas right, the pain of finding them wrong. Out of these can develop more complex states. Thus, success can breed overconfidence and arrogance, and challenges from other scientists can elicit defensiveness or anger. In susceptible individuals these emotions can sway judgment, overcome objectivity, and lead to exaggerated or false claims.

Entangled with the emotional elements are the almost universal aspirations to security and career advancement, occasionally sharpened by hunger for fame and fortune. Those urges can motivate a scientist to cut ethical corners in order to satisfy a need, for instance, for one more confirmatory experiment to wrap up a thesis or a research project; for a few more publications to beef up a job application or a recommendation for promotion; for a critical result to support a grant application or a claim to priority in an important discovery. Many of these motives can be glimpsed in the articles that accompany this chapter.

In the first article a distinguished biochemist, Efraim Racker, recounts his personal experience as a victim of what he calls "professional" fraud, an extreme form of scientific misconduct meticulously planned and executed by highly intelligent individuals who, Racker believes, are mentally imbalanced and among other obscure motives may harbor an unconscious wish to be caught. Several prominent laboratories have been targets of such attacks in recent decades. Although much less common than "amateur" fraud, incidents of professional fraud are likely to recur sporadically wherever research is conducted, since they seem to be driven by sources within a disturbed mind rather than by external circumstances.

The next two articles summarize some of the main events of the so-called Baltimore case. This has become the cause célèbre among modern cases of scientific misconduct, one that can serve as a cautionary tale of how *not* to deal with scientific misconduct and the whistleblowers who seek to expose it. The subject of countless informal and formal inquiries up to the congressional level, and of a multitude of articles, books, open letters, and even plays, this case after a decade is still not fully resolved. Philip Hilts, a journalist, under the dramatic title of "The Science Mob," describes in one of the articles how scientific fraud and its handling or mishandling by the scientific establishment have emerged during recent years as pervasive problems. As a prime exhibit he presents the Baltimore case. The other article deals with the esoteric details of the alleged scientific misdeeds at the root of that case. Some of those details remain uncertain to this day, but the ineptness of individual and institutional actions surrounding the case has been amply documented and has left a long trail of controversy, bitterness, and disrupted careers.

In contrast, the cases described in the last article, both of which occurred in the same laboratory at Cal Tech, have been regarded as models of deft handling of suspected scientific misconduct and fraud. Despite the effectiveness of the procedures, however, in these as in almost all such cases troubling questions remain about the intentions and motives of the accused, about the capacity and responsibility of scientists to police each other, and about the fairness and impartiality of investigations carried out by institutions whose own reputations are at stake. These are important questions, well worth pondering for their own sakes.

A View of Misconduct in Science

EFRAIM RACKER

In the United States, suspicion of fraud has become a cancer in research. To tackle the problem, real and perceived, scientists will have to become more practiced at public relations.

In this article I give a personal view of misconduct in science. It is based on one experience in my own laboratory and three others in different laboratories, and I shall consider four aspects of the problem: the scientists who were involved; the supervisors and collaborators; the university and funding agencies; and the government as represented by Congress.

THE SCIENTISTS WHO WERE INVOLVED

I have had direct or indirect contact with four young scientists who were accused of fraud. The first worked in the laboratory of Dr. Carl Cori at Washington University in St. Louis; the second in the laboratory of Dr. David Green in Madison; and the third in the laboratory of Dr. Melvin Simpson at Yale University and later with Dr. Fritz Lipmann at Rockefeller University. (Cori, Simpson, and Lipmann all published retractions. Green retracted the work with which he was concerned at a public meeting of the Federation of American Societies for Experimental Biology.)

Reprinted by permission from *Nature* 339(1989):91–93.

Finally there is Mark Spector, who worked in my laboratory, and most of this article will discuss his case.

What did these scientists have in common? They had an outstanding intellect; they were all informed in their research field and recognized the missing links that if solved would represent a breakthrough. They were skilled experimenters and joined laboratories that were in the forefront of a particular research field and that were headed by an investigator with a reputation of integrity. As far as I know, none of them ever admitted being guilty. These are trademarks of the "professionals" and are not universal. "Amateurs," who for example publish data on patients that do not exist, are usually more readily discovered by colleagues and students. The spiritual damage caused by scientific fraud is irreversible, and those involved are and should be reported and prosecuted irrespective of whether financial losses are involved.

One evening in 1972 (the year of Nixon's Watergate) I received a phone call at home from a man in Washington, D.C., who told me he was writing a book on fraud in science. I forgot his name and I don't know whether the book was ever published. He told me he knew that I was familiar with several scientists who had been accused of fraud, and that he had found out about a sin of my youth—an interest in psychiatry. He challenged me to give him a psychological analysis of these men. I briefly outlined to him the common traits mentioned above. Because these three men were very smart, I pointed out, they must have known that in view of the excitement caused by their claims the work would be repeated promptly and if there was fraud it would be quickly detected. I concluded, therefore, that they were mentally unbalanced and subconsciously wanted to be caught. The man from Washington responded, "You mean, like Richard Nixon?"

In my opinion scientific fraud perpetrated by what I call the "professional" springs from an unbalanced mind. Perhaps as with most professional criminals they are emotionally and mentally ill, often seeking self-destruction. But I must stress, before being misunderstood, that such a view should have no bearing on our verdicts of guilt and punishment. In most cases psychiatry has no place in the courtroom unless it becomes more scientific than it is today. This view is not generally accepted and is incompatible with criminal prosecution according to current laws. Scientists who commit fraud are rarely psychotic and can and should be prosecuted.

Mark Spector, as I knew him, was a brilliant young man. He wrote poetry, painted, and was a superb experimenter. A young scientist who watched him perform experiments at the National Institutes of Health commented that it was like watching Beethoven play the piano (I hope that the NIH scientist had a better knowledge of biochemistry than of musical history). Spector's lectures were spellbinders, and as a graduate student he received lecture invitations from many places, including NIH. He was also an excellent teacher, and one of my other graduate students is still grateful for what Mark had taught him. Outside pressures can drive investigators to fraud, but it was not a factor in the four cases I knew. Spector worked long hours and late into the night. I told him several times to slow down, assuring him that with his skill and teaching talents he would get a faculty position at a very good university. Whatever pressure there was came from within.

When Mark came to my laboratory we were working on the role of the $(Na^+ + K^+)$ATPase in glycolysis of Ehrlich ascites tumor cells. I also had several postdoctoral fellows working on protein kinases in mammalian and plant cells. Based on experiments he claimed to have done, Spector proposed that a cascade of several protein tyrosine kinases was responsible for the phosphorylation of the $(Na^+ + K^+)$ATPase and its inefficiency of operation in Ehrlich ascites tumor cells, a clever concept, compatible with our ideas on the mechanism of pump inefficiency.

Spector generated data by substituting radioactive iodine, which is widely used to label proteins, for radioactive ATP. He showed me the protocols and autoradiograms day by day. I suggested experiments: As usual, some of them worked, others didn't. He was always downcast when they didn't. He was invited by Dr. Edward Scolnick at NIH to demonstrate some of his experiments, the first of which was a complete failure. He called me in desperation, suggesting that the enzymes might not have survived the trip. He told me where he stored them, and asked me to ship them by Federal Express. With the enzymes I sent him, the experiments worked. Needless to say, he must have had the iodinating reagents with him all the time. Dr. Scolnick told me much later that the samples Spector left behind contained iodine; Mark was an artist of deception.

It was sheer luck that we discovered what he had done. We collaborated with Dr. Volker Vogt, then an assistant professor at Cornell, who on examination of one of Mark's gels accidentally covered it with a glass plate and noted that the monitor still registered counts. He realized that the glass should have blocked the radiation emerging from radioactive phosphate but not those from iodine, and reported the observation to me.

Another way of deception that scientists should be aware of is the simple method of spiking. Spector claimed that he had discovered a new ATPase activity in a cytoskeleton preparation from Ehrlich ascites tumor cells. It was exactly what I wanted to hear and I myself started to analyze and purify the enzyme. Because Mark collected the cells for his other work, he just turned the triton-insoluble fraction over to me to keep me busy. After three weeks of intensive bench work I became concerned, because the properties of the enzyme clearly resembled a well-known viral ATPase that we had studied. Sure enough, with electron microscopy we detected a few virus particles in the preparation. Soon thereafter Mark's preparations became weakly active or inactive. I thought that some of our mice may have been infected by a virus and dropped the project. After the iodine incident, I realized that he could have spiked the cytoskeleton preparation with samples of the AMV virus which I had stored in my deep freeze. Mark knew where everything was kept.

One of Mark's claims was that he had isolated a new protein kinase activator with properties similar to a transforming growth factor described by Dr. George Todaro. We received a sample of this factor from Dr. Todaro, Mark was excited; it worked in his assay. Dr. Todaro came to our laboratory and repeated the experiments with the enzymes he was given. He returned to NIH with a sample of Mark's purified activator and, at the next meeting of the Federation of American Societies for Experimental Biology, reported that Mark's factor was active as a growth factor. I suspect that what he tested was Mark's activator spiked with the growth factor Dr. Todaro had generously given to us. There was a similar spiking incident with an antibody he had sent to Dr. Ray Erikson. Dr. Edwin Krebs was excited by our findings and sent his collaborator Dr. Linda Pike to work with Mark. Needless to say, the data were spectacular and we were ready to publish when the iodine surfaced. Dr. Pike, herself a talented experimenter, was duly impressed by Mark's skillful experimentation.

I am reporting these incidents here in some detail and for the first time, mainly to make a point that I shall reemphasize later—it is difficult to catch a "professional." I shall describe in the next section how we responded.

THE SUPERVISORS AND COLLABORATORS

Supervision. Careful supervision of technicians, students and postdoctoral fellows is essential—not because of fear of fraud, but because it is part of our teaching responsibility to detect errors, inappropriate controls and so on. It has been a long-standing practice in my laboratory that important discoveries are repeated either by myself or by other colleagues. The reasons for that practice are that it detects experimental errors, it enhances confidence and it greatly helps in the write-up to ensure duplication in other laboratories. Some of my own experiments are duplicated by others in my laboratory. Many of Mark's experiments were repeated by me with enzymes given to me by him, and his experiments on the activator were duplicated by Dr. Todaro. One possible explanation is that the enzymes he gave us were spiked—for example with the cAMP-dependent protein kinase, the activator preparation with cAMP—but I have no proof of this.

I would like to describe in some detail one particular incident that convinced me at the time that Mark's discoveries were genuine. At an early stage he showed me data that documented the phosphorylation of the β-subunit of an $(Na^+ + K^+)$ATPase isolated from Ehrlich ascites tumor cells by an endogenous protein kinase. It was exactly what I had hoped for and I thought it was too good to be true. I asked a postdoctoral fellow in my laboratory to repeat the experiments and to explore other $(Na^+ + K^+)$ ATPase preparations that I stored in my deep freeze. Among them was an $(Na^+ + K^+)$ATPase from *Torpedo californica,* which has a β-subunit with a different electrophoretic mobility to that of the mouse enzyme. A postdoctoral fellow reported to me that she could not repeat Spector's experiments. Mark was called in, quizzed her and quickly discovered that, by not freezing his enzymes in liquid nitrogen, she had failed to follow his instructions exactly. The experiment was repeated with fresh solutions and it was a remarkable success. All ATPases were phosphorylated on the β-subunit, including the *Torpedo* preparation with its more rapidly moving β-subunit. In later years I have shown this autoradiogram to many visiting experts asking how it could have been faked. Nobody could understand it. It was only when I wrote this article that I thought of an explanation that is staggering—during the night Mark could have found her protocol as well as the ATPase preparations in the deep freeze, labeled them with iodine, and placed them on an acrylamide gel which he substituted for hers.

Response to Suspected Fraud. The day after Vogt informed me that he had discovered iodine in Mark's samples we confronted him, demanding an explanation. He was quite calm, denied any wrongdoing, and claimed that somebody else must have put the iodine into his samples. I felt it was important not to announce a premature verdict, and therefore gave him three weeks to prepare fresh enzymes, which he was to turn over to me for assay with my own reagents. When he failed to come through, I notified the graduate school to withhold his doctoral degree for which all requirements had been fulfilled. I withdrew two papers in press and wrote a letter to inform NIH. I also sent a letter of retraction to *Science* (213:1313. 1981) in which a review paper of our work had previously been published. Because some of Mark's experiments had been repeated by myself and others, several postdoctoral associates later attempted to determine what could be salvaged. We published several papers that were based on Mark's findings but there were enough differences to discredit all of his claims.

UNIVERSITY AND FUNDING AGENCIES

How should these institutions handle the prosecution? Some of our faculty members felt that Mark should be taken to court and punished. I decided not to recommend this route to Cornell University, for the following reasons. Once again, Mark never admitted that he had committed fraud, and I felt that the evidence of one experiment at Cornell and one at NIH in which radioactive iodine was found instead of radioactive phosphorus could well not have convinced a jury or judge. Mark would have made an eloquent and persuasive witness. He could not have denied the presence of iodine but could have denied that he had added it. Had we lost the case, Mark would have demanded his PhD degree and would have received it. As it was, he refused my request to resign voluntarily and threatened a lawsuit. I talked to Mark's mother, a reasonable lady with a great deal of insight, who finally persuaded him to withdraw without further action.

This brings me back to prosecution. Irrespective of causes, society must punish fraud. Although I don't know what eventually happened to Mark, his scientific career was destroyed and he gained no money to flee to South America. Such punishment is severe. In fact, I received quite a few letters accusing me of dealing too harshly with him and failing to give this brilliant young man a second chance. As far as I was concerned this brilliant young man was ill and there was no cure for his illness. Because I felt this way, I was never angry with Mark but was not willing to be lenient either. (The young scientist from Cori's laboratory was given a second chance and became a professor at the School of Public Health of Johns Hopkins University. He published several sensational papers that were never verified, and died at a young age.)

It seems obvious that each case must be handled individually. When the evidence can stand up in court, conviction and jail sentences are in order. There may be mitigating circumstances, such as undue pressure and confession, which may warrant less severe measures—loss of reputation and a job is in itself severe punishment.

THE GOVERNMENT AND CONGRESS

I think that scientists should take the stand that fraud, when established, is a matter for the courts and we should show that we are serious about it. The obligation to meet legal standards as in other cases of fraud would discourage pursuit of trivial cases that could be settled within the university or by NIH. It is sometimes impossible to distinguish between honest mistakes and fraud. We must try to educate the public and members of Congress that there is little to gain and a great deal to lose by involving Congress in scientific fraud. But Congress rightly insists that universities and funding agencies must forcefully investigate cases where fraud is suspected. We have to convince members of Congress that the operation of science is similar to that of self-defrosting refrigerators, that the occasional ice of fraud is melted away by the heat of progress.

Audits will have little effect on fraud and the expense is not justified. The argument of Dr. Drummond Rennie, a deputy editor of the *Journal of the American Medical Association,* that we must persuade Congress not to get involved in audits, is valid, but I disagree on how to do this. He recommends that we scientists should undertake the task, but this would be a grave mistake. At best audits will provide us with numbers that show that fraud is relatively rare, but I doubt that our politicians will accept such a verdict. Audits may discover misconduct in clinical trials that would probably come to light anyhow. But they will not uncover the "professionals." Let anyone look at Mark Spector's autoradiogram and at the protocols of the experiments with spiked enzyme preparation and try to discover fraud.

The threat of audits will not scare those intent on fraud; it will only increase their sophistication in keeping notebooks, and they will use symbolic gloves that will prevent the detection of the fingerprints of their deceit. Audits may pick up discrepancies, bad controls and inappropriate authorships, but it will be very difficult to differentiate between fraud, poor judgment, and errors in interpretation. I think historians could find such errors in the papers of our greatest scientists. Let them study the history of glycolysis and oxidative phosphorylation—can we call the mistakes or errors of judgment made by Louis Pasteur, Otto Warburg, or Peter Mitchell scientific misconduct?

Inappropriate authorship is a common and unacceptable practice that can be reduced without audits. Editors of journals can specify their requirements and demand a list of what each author has contributed, signed by all authors. Painful, yes, but better than audits. I am concerned that if scientists start to audit data of their colleagues, the next step may be audit of experiments. All this will be humiliating, expensive, ineffective, and will not significantly change the overall picture of misconduct. If Congress insists, let it conduct audits by its own staff and discover what audits are worth.

We have to educate the public, the press, and members of Congress that if there is any human endeavor in which crime doesn't pay, it is in science. We have to convince them by documentation that science has made enormous progress in the past few decades and that the few cases of fraud did not slow it down, but that government interference and their lack of enthusiasm for science will. We have friends in Congress; we must ask them to help us. Congress has already done its job of persuading scientists and administrators to be more vigilant and to pay more attention to cases of alleged fraud. We have a crisis on our hands, but should not overreact.

Our task is to convince members of Congress that science is too fragile to be used as a political football. Most scientists of today can take the heat, but what about the next generation? Our students are already dismayed by the prospect of fighting for grants and university positions, and are turning to other and more lucrative professions. Let us collect data and carry out interviews with students that will document this point. If young scientists are also threatened by government investigations of un-American and unscientific conduct, they will simply stay away from a scientific career.

To conclude, I have a few recommendations. We should present to Congress a brief yet precise plan of what we can do, and we need a science lobby that will prepare and present the plan. I don't know all that such a manifesto should include, but I have some suggestions. It should define a role for the universities and the departmental chairmen, to keep an eye out for irregularities as well as for undue pressures on students and faculty members. It should include a role for editors in dealing with improper authorships and for setting higher standards for the reviewing processes. Faulty citation of references is a form of scientific misconduct that is easily criticized but often innocent. Reviewers should be requested to take an active role in pointing out such omissions. Editors who do a sloppy and erroneous job of reviewing are also guilty of scientific misconduct and should be eliminated from editorial boards. Just as members of Congress, editors and reviewers of manuscripts and of grant applications should be accountable to a committee on ethics. We should plan to educate practicing scientists on how to supervise better the work of students, not as a device to prevent fraud but because that is a responsibility of any teacher.

We should also convince the press that they need to help us in informing the public of what science is all about. I wish we could persuade Congress to think of curbing inaccurate and sensational journalistic versions of the truth; there must be freedom of the press but without the freedom to distort and invent. Members of Congress, I trust, are well aware of this problem. All this will require time and patience, as well as vigilance—remember how many decades it took to convince even a fraction of the public of the simple fact that smoking is hazardous to health? Education is a slow process and imaginative ideas to accelerate it are badly needed.

The Science Mob: The David Baltimore Case—And Its Lessons

PHILIP J. HILTS

> NEOPTOLEMUS: You're capable, Odysseus, and resourceful. But you have no values.
>
> ODYSSEUS: And where's the value in your carrying-on?
>
> NEOPTOLEMUS: Candor before canniness. In doing the right thing and not just saying it.
>
> —Seamus Heaney's translation of Sophocles' *Philoctetes*

In the years before World War II, science was a small, charmed profession. In 1940 there were about 200,000 scientists and $70 million in federal money. Scientists were a contemplative order, and their exposure to the world was limited. When an occasional question of sloppiness or misconduct arose, it was quietly resolved within the confines of the profession. But now, as the number of scientists reaches 1 million and their share of the nation's federal budget reaches $25 billion, the demands for greater accountability and openness are understandably more insistent. Though scientists would like to remain aloof, a brotherhood whose standards and integrity remain above public reproach, that era is over.

Stories about scientific misconduct are no longer an aberration. Indeed, in recent years the most notorious cases have involved some of this country's most reputable scientists and universities. In 1983 John Darsee, a researcher at Harvard Medical School, was found by the National

Reprinted with permission from *The New Republic* (May 18, 1992):25, 28–31.

Institutes of Health to have faked some data in his studies on heart disease. In 1984 the National Institute of Mental Health concluded that Stephen Breuning, a researcher at the University of Pittsburgh, had fabricated data in a paper about drug therapy for hyperactive children. In both cases earlier internal university investigations had cleared the scientists of blame. Robert Gallo, the chief of the Laboratory of Tumor Cell Biology at the NIH and the co-discoverer of the cause of AIDS, is under investigation by several federal agencies for not giving sufficient credit in 1984 for work performed by French scientists. A recent NIH report found him not guilty of misconduct but detailed several instances of irresponsible behavior.

The NIH, which is charged with investigating allegations of misconduct in federally funded research at universities, examines a few dozen such cases each year. It is impossible to say how many others remain under wraps at the universities. The reluctance of administrators to root out cases of misconduct by faculty is hardly surprising: when one comes to the attention of federal investigators, and the perpetrator is found guilty, his federal research funds are withdrawn. More important in a system in which reputation is paramount, a charge of misbehavior represents a permanent disgrace—a lingering impediment to future federal and private funding.

Perhaps the most remarkable case of misconduct in the annals of American science is the one known as "the Baltimore case." The most protracted scandal of the last several years, it stands as the exemplar of what's wrong with the defensive and self-regulating structure of the American scientific establishment. It's named after the scientist who refused to investigate allegations of faked notebooks, Dr. David Baltimore, rather than after Dr. Thereza Imanishi-Kari, the scientist charged with the fraud. And it's famous less because of the nature of the fraud than because Baltimore himself determined to make it famous. Like a Greek tragedy, it turns on a character flaw in the protagonist, unseen by himself but excruciatingly obvious to the audience, that allows him to commit a sequence of improbably foolish acts. Each leads to the final—maddeningly avoidable—fall.

The case is quite simple in many respects, and it could have been quickly resolved at the start. Instead it has dragged on for the past six years, involving dozens of eminent scientists who rallied behind Baltimore, and provoking two university inquiries, two formal investigations by the NIH, and three congressional hearings by the oversight committee responsible for looking into government fraud. And still it is not over. The NIH has not yet finished its investigation, and a grand jury in Baltimore is considering indictments against Imanishi-Kari.

What we now have, though, is a thorough draft report by the Office of Scientific Integrity at the NIH that provides a factual guide to the impenetrable. From this and the testimony of each side since the draft was leaked to the press last spring, we know at least the sequence of events that led to the public humiliation of Baltimore, a Nobel Prize winner and former head of the Whitehead Institute and president of Rockefeller University (Baltimore was finally pressured to resign from Rockefeller last fall by senior faculty who felt the ongoing scandal was an embarrassment to the university). We cannot say why Baltimore did what he did. I have asked him repeatedly, and he is unable to say why.

The case began with a research paper, published in the journal *Cell* on April 25, 1986, titled, "Altered Repertoire of Endogenous Immunoglobulin Gene Expression in Transgenic Mice Containing a Rearranged Mu Heavy Chain Gene." The paper, written by Imanishi-Kari and co-authored by Baltimore (then her colleague at MIT) and three other scientists, described experiments that purported to show that when scientists inserted a foreign gene into mice, it did not, as expected, just make foreign antibodies. Rather, it had some unknown effect on the mouse's own genes, altering them to include antibodies that mimicked the foreign antibody. The paper implies that it might sometime be possible to gain command of the body's defenses by introducing foreign genes that would recruit the natural ones to attack a selected target.

The paper began to unravel almost immediately, even before publication. The warning signs came from the MIT postdoctoral student, Dr. Margot O'Toole, assigned by Imanishi-Kari to extend the work to the next step. She could not duplicate the work and wasted almost a year demonstrating that important experiments in the paper were wrong. It is always dangerous for postdoctoral students to challenge their superiors, upon whom they rely for every detail of their professional life, including money, lab space, and the opportunity to publish. This particular challenge would require either an extensive correction or a withdrawal of the paper, an unusual procedure that would embarrass all of the authors.

In May 1986 O'Toole first took the uncomfortable facts to her thesis adviser and two other scientists at Tufts University, which was about to hire Imanishi-Kari. They were concerned enough to call in Imanishi-Kari for proof of the work she'd done, but after a quick perusal of several pages of her notes on the experiment, they decided that whatever problems existed need not be disclosed. (Forensic experts at the Secret Service now say two of the pages of evidence she brought were fabricated just before the meeting. Tufts hired Imanishi-Kari, where she remains today as an assistant professor in the Department of Pathology.)

O'Toole then went to the dean at MIT, who asked Dr. Herman Eisen, a friend of Baltimore's, to look into the case. Though Eisen was the officially designated investigator at MIT, he never looked at Imanishi-Kari's lab data or her notes. He did not question Imanishi-Kari, O'Toole, or Baltimore. Instead, he quickly read a memo from O'Toole on what was wrong, discussed the matter with the Tufts scientists, and later wrote a report saying that there appeared to be errors in the *Cell* paper and differences in interpretation between Imanishi-Kari and O'Toole, but that this was "the stuff of science," and not misconduct. (Several months ago, in a meeting with scientists at Harvard who continued to be perturbed by the case, Eisen admitted that he did not read O'Toole's memo carefully. He also said he "never believed" the theory behind the part of the paper done by Imanishi-Kari, and so was not particularly concerned with the accuracy of the evidence itself. Such rationalizations could hardly have provided the reassurance the group was looking for.)

Finally, on June 16, 1986, O'Toole herself confronted Baltimore and Imanishi-Kari at a meeting also attended by Eisen and another co-author of the *Cell* paper, David Weaver, a member of Baltimore's lab. She was the only one who brought data to the meeting—17 pages from Imanishi-Kari's notes. (Investigators at NIH later said those pages were prima facie evidence of trouble because they showed results opposite from those reported in the paper.) According to O'Toole, Imanishi-Kari admitted at the time what she has come to state publicly: some of the work cited in the paper was not done, and other work got different results than what was reported. At the end of the meeting, O'Toole asked that the paper be corrected or withdrawn. Baltimore replied that such problems with accuracy are not unusual and they need not be corrected—a startling new standard for scientific inquiry.

He said that the scientific process is "self-correcting"—meaning that other scientists will eventually figure out that the published work was wrong. It is true that honest work is often wrong and requires another study to reveal that. But Baltimore was extending the notion of self-correction to cover errors he knew existed but decided not to report. Thus he was dooming some scientist to repeating work that need not be repeated, merely to maintain his own unblemished record.

O'Toole pressed him. He says he told her she could write to *Cell,* but that if she did, he would write his own letter endorsing the paper's results, and that he couldn't imagine they would accept her letter then.

O'Toole says that she left the meeting feeling beleaguered and decided to let the matter drop.

However, by July 1986 the case was sniffed out by a pair of self-appointed fraud scouts at NIH, Walter Stewart and Ned Feder. They had heard of the case through the grapevine and began to press O'Toole to give them information about it. Though they have no official status as investigators, the burden of pressing such cases went to them because they were willing to do the work necessary. There is in fact nobody in science directly assigned to study and adjudicate potential cases of misconduct. They also alerted Representative John Dingell, chairman of the House Subcommittee on Oversight and Investigations, who oversees the workings and misworkings of federal agencies. He began his own prolonged inquiry and eventually held two hearings on the case, one in April 1988, the other in April 1989.

In January 1988 Stewart and Feder's work and Dingell's investigation finally prompted the NIH to appoint an official committee to investigate the matter. But at first, and true to form in investigations carried out by scientists, the NIH put two of Baltimore's close associates on the panel, Frederick Alt of Columbia, a co-author with Baltimore on more than a dozen papers, and James Darnell of Rockefeller, co-author on Baltimore's very successful textbook on molecular biology. The third panel member, Ursula Storb of the University of Chicago, was later found to have written a letter of recommendation for Imanishi-Kari.

That summer Baltimore began a national campaign designed to derail the NIH and congressional investigations. He attacked O'Toole as a "discontented post-doc" in a letter to the NIH, and he and several friends at MIT orchestrated the writing of letters to more than 400 colleagues in which the investigations were declared to be a threat to science itself. Baltimore at the time was chief of the Whitehead Institute, MIT's molecular biology research institute, as well as a professor at MIT, and he committed tens of thousands of dollars of the institute's money to lobbying, including the hiring of Akin Gump, a high-priced Washington law firm, to press his arguments upon Congress.

Baltimore cast the conflict as one of outsiders invading the sanctuary of science. They were, he said maliciously misrepresenting a scientific dispute about error as a case of fraud. He appealed to the xenophobia of other researchers in asking them to rally round him. In one letter, a close friend of Baltimore's, MIT's Phillip Sharp, urged his colleagues to write op-ed pieces, and letters to the editor and to Congress. His sample letter to Congress said: "I believe that to continue what many of us perceive to be a vendetta against honest scientists will cost our society dearly. If scientists who have been exonerated of all wrongdoing must continue to defend themselves against vague and shifting charges, all members of the scientific community must be afraid." Robert E. Pollack, dean of Columbia College, did write an op-ed piece in the *New York Times* in which he deplored congressional meddling in science: "The way Dr. Baltimore is being treated means that witch-hunts are in the offing," Pollack declared. "If Congress legislates against error in science, there is no chance that a sensible young person will choose to be a scientist." The number of combatants in the fray grew, until half a dozen Nobel Prize winners and eminent scientists from Stanford, MIT, Harvard, Tufts, and Rockefeller had taken up the cudgels. Baltimore and his lobbyists arranged for a bevy of distinguished scientists to go to Washington on his behalf. They had seats reserved just behind Baltimore at Dingell's second congressional hearing in April 1989, facing Dingell.

David Baltimore was the only source of his colleagues' certainty that the case was one of error and not fraud. But Baltimore himself had not looked at the evidence in detail; in fact, he said it was not his business to look at it. What he did know, at the very least, was that there were false statements in the paper. For example, one of the problems raised in the summer of 1986 was that one of the reagents did not perform as stated in the paper. That September, several months after Eisen had concluded his inquiry into the matter, Baltimore wrote in a letter to him (made public under subpoena): "The evidence that the Bet-1 antibody doesn't do as described in the paper is clear. Thereza's statement to you that she knew it all the time is a remarkable admission of guilt . . . Why Thereza chose to use the data and to mislead both of us and those who read the paper is beyond me." More interesting, a few lines later Baltimore admitted choosing to mislead those who read the paper, and he gave a reason why. "All authors do have to take responsibility for a manuscript, so all of us are in a sense culpable, but I would hate to see David's [David Weaver] integrity questioned for something he accepted in good faith . . . The literature is full of bits and pieces now known to be wrong, but it is not the tradition to point each one out publicly."

He said that no correction should be published but that he would privately let others know that Imanishi-Kari's data "are not reliable, and I, for one, will be skeptical of Thereza's work in the future." Later Baltimore told the Office of Scientific Integrity that he was not proud of this letter and his decision to advise against a correction and added, implausibly, that probably he and Eisen had misunderstood Imanishi-Kari's explanation of her misdeed. Imanishi-Kari is originally from Brazil and has a mild accent.

When Dingell subpoenaed Imanishi-Kari's notebooks in preparation for the congressional hearings in the spring of 1989, she met with Baltimore and his lawyer Normand Smith. She confided that she really had no notebooks, only loose sheets of paper, spiral-bound pads, and folders. Researchers' notebooks often are not pristine, but when subject to examination they must make sense. What should I do with this mess? she asked. Either Baltimore or Smith—neither will be definite about it—told her to assemble them into a notebook.

On April 25 Dingell's staff invited Baltimore in for a private talk. It was nine days before the hearings were to take place. Dingell's staff had taken the notebooks to the top forensic experts at the Secret Service, who reported that all the signs of outright fraud were there. Dingell's staff felt that if Baltimore got a look at this new data, he might have a chance to regroup, back away, and offer to help resolve the matter. He was told that the Secret Service had found that 20 percent of Imanishi-Kari's notebook material showed evidence of being faked. But Baltimore still didn't back down. In fact, at the hearings he was asked how Imanishi-Kari came to make the notebooks. He replied that he did not know.

The paper and typefaces from mechanical data counters did not match those used in the lab in 1985 when the data was supposed to have been taken. Rather, all the signs matched perfectly data from another time in the lab—several years before, when it would have been impossible for the experiments to have been done. The paper on which the purported data was recorded was a peculiar shade of yellow-green, unlike anything seen in the lab for years. And, astonishingly, a number in one of the notebooks was changed, simply whited out both front and back. Dates in Imanishi-Kari's notebook pages were out of order, overwritten, and some were clearly wrong for the experiments represented on the page. Later, when confronted with these by the NIH investigators, Imanishi-Kari said that dates "don't mean anything." Maybe they are not even dates, just numbers. Numbers referring to what? she was asked. "I don't know," she said.

Baltimore was clearly shaken by the meeting. Those present said his color sank, and they feared he would be sick on the spot. But his recovery was quick. In a subsequent meeting that must be considered at the least highly improper, he met with the NIH investigators and with Imanishi-Kari to talk about the testimony they would give before Dingell.

For example, when Imanishi-Kari suggested the paper may have gotten discolored by leaving it in the sun, NIH investigator Dr. Hugh McDevitt said that story would not work because they already knew it was not true. He offered the possibility that there was another explanation, one she hadn't suggested yet.

When it came time to testify, Baltimore delivered as remarkable a piece of oratory as a scientist ever did before Congress. "The Secret Service apparently conducted a nine-month forensic analysis of Dr. Imanishi-Kari's laboratory notes," he said. "In a charade of helpfulness, they presented a partial oral summary of their findings on Tuesday, April 25. That presentation was designed to terrify without providing any substance . . . last Sunday, some written materials were provided. And based on those and what I have heard today, there is still nothing from the Secret Service investigation that causes me to doubt the validity of the *Cell* paper." Though Baltimore himself had almost single-handedly created the whole spectacle, he went on to chastise Dingell. "I must tell you, Mr. Chairman, I am very troubled about how this situation got so out of hand. I have a very real concern that American science can easily become the victim of this kind of government inquiry . . . Professor Imanishi-Kari is also a victim . . . She deserves my support, and the support of all scientists, for any of them could be in her shoes."

No one doubts that Baltimore is a brilliant scientist. But those who know him have seen another, more childish David Baltimore in outbursts from time to time. His extraordinary success may also have led him to feel invulnerable—able to deflect personal scandal merely by bringing the weight of his reputation to bear. From his weakness we see the weakness of science: that it is a human enterprise. Its practitioners struggle always against emotion and prejudice, and never fully overcome them.

O'Toole's plight illustrates the dangers in a hierarchical system where a scientist is inaudible to all those above her rank. When she made her charges, the senior scientists turned and spoke to one another. Eisen talked to Baltimore, Tufts to MIT. Later, when Stewart, Feder, and Dingell joined in, they likewise carried no particular status in science. Baltimore and others even chose to contradict the forensic experts at the Secret Service, who surely know their business.

O'Toole, who is now working at the Genetics Institute in Cambridge, Massachusetts, after a long hiatus in which no one in the field would hire her, believes that the only way to avoid another Baltimore case is to have the investigations of such matters open and public. Other scientists have had a similar response. Dr. Walter Gilbert, a Nobel Prize winner in molecular biology from Harvard, says: "Some of us are just aghast at David's behavior. Through his own doing, the case became a dramatic test of power between the Congress and the scientific establishment. It became a case of how science should be supported and reviewed. He tried to make it a test case, rather than say, 'I'm sorry,' and walk away, or, as any scientist should say that if the work was wrong he would be responsible and withdraw it." The case, Gilbert says, has proved to be a healthy reminder to scientists "that lab notebooks are open documents, that all the authors on a paper are responsible for it. Fact-finding must be done vigorously and impartially, rather than by the friends of the person involved. What has not been healthy is the failure of the institutions—both the universities and the NIH—to investigate quickly and thoroughly."

But the Baltimore case echoes something deeper in the scientific world than mere secretive procedures and mutual, collegial protection. It reveals something about the nature of the scientific mind itself. The key to science, the physicist Richard Feynman wrote, is "a kind of scientific integrity, a principle of scientific thought that corresponds to a kind of utter honesty—a kind of leaning over backwards. For example, if you're doing an experiment, you should report everything that you think might make it invalid—not only what you think is right about it." These are exacting standards, and ones that human beings—with all their propensity for pride, vanity, and ambition—regularly fail to live by. For too long scientists—and the society that supports them—have believed that they are somehow immune to these imperfections, that their professional integrity should therefore be placed beyond the troubling, open, sometimes misplaced scrutiny of a liberal democracy. The last few years should prove beyond any doubt that those scientists are all too human and that such scrutiny is all too often merited.

David Baltimore clearly failed as a scientist—through his carelessness, his willful oversight, and his extraordinary attempts to protect his own reputation at the expense of a conscientious young colleague. In the end, Baltimore inadvertently revealed just how vulnerable the scientific profession is to abuse by those entrusted to protect it.

Verdict in Sight in the "Baltimore Case"

DAVID P. HAMILTON

Congressional and NIH investigators have been sifting through forensic analyses of Thereza Imanishi-Kari's notebooks for nearly 2 years; at least two sets of data are in doubt.

When a team of researchers published a paper in *Cell* in April 1986,[1] they probably expected to generate scientific controversy, for they were reporting results that challenged the conventional wisdom about how immune responses are regulated. But they surely didn't expect the attention to which their paper has been subjected over the past 5 years: two university reviews, a congressional inquiry, and two investigations by the National Institutes of Health.

What began as a laboratory dispute between Tufts immunologist Thereza Imanishi-Kari and her then postdoc, Margot O'Toole, has escalated into one of the more celebrated cases of alleged scientific fraud in years. Congressional investigators working for Representative John Dingell (D-MI) have privately accused coauthor and Nobel laureate David Baltimore of using his prominence to "cover up" error and possible fraud; defenders of the paper's authors have depicted Dingell's inquiry as a political assault on the foundation of science. No matter whose side you've been on, the affair has been an ugly one. Rightly or wrongly, everyone has been impugned: the authors for, at best, doing shoddy science; O'Toole for blowing errors out of proportion; the congressional investigators—who have been assisted by unofficial NIH fraudbusters Walter Stewart and Ned Feder—for conducting a witch hunt; and the official investigative bodies, including Tufts University and the Massachusetts Institute of Technology, for carrying out a series of botched inquiries that had the effect of derailing O'Toole's career. And the accused have all been left in a form of purgatory, their reputations tarnished for years, with neither final condemnation nor vindication. Until now, perhaps.

After 22 months of investigation, an NIH panel is expected to deliver its verdict in the case within the next few weeks, assuming that a recent court challenge to NIH's Office of Scientific Integrity (OSI) doesn't derail the investigation (*Science,* 11 January, p.152). A draft reportedly is circulating among the committee members and will eventually be sent privately to all parties concerned. Once the OSI has considered any comments the authors and institutions may wish to make, the report will be made public. Although neither NIH panel members nor OSI officials will comment publicly on the ongoing investigation, *Science* has assembled an account of the main issues with which the NIH panel is wrestling, based on a months-long examination of testimony given at two bruising public hearings by Dingell's subcommittee, additional unpublished documents, and in-depth interviews with Imanishi-Kari, congressional investigators, several independent immunologists, and sources close to the NIH investigation. (Baltimore declined to be interviewed for this article, pointing out through a spokesman that he has not been accused of misconduct and is not a target of the investigation.)

The central charge facing the NIH committee is the one raised by O'Toole almost five years ago; that Imanishi-Kari's original laboratory data do not support the authors' published contention that a gene transplanted into a line of mice indirectly changed the repertoire of antibodies produced by the mouse's own genes—the *Cell* paper's main thesis. The evidence available to *Science*—especially Secret Service forensic analyses of ink, paper, and dates in Imanishi-Kari's laboratory notebooks, as well as of paper tapes produced by radioactivity counters—casts doubt on the authenticity of one key set of data. And the recent emergence of contradictory original data in a grant application raises questions about a second.

Investigative panels at MIT and Tufts found O'Toole's charges unwarranted in 1986, and an NIH panel concluded in February 1989 that there was "no evidence of fraud, manipulation or misrepresentation of data." But when Dingell scheduled hearings in May 1989 to present forensic evidence in the case developed by the Secret Service, then NIH director James Wyngaarden added two new members to the original panel,[2] reopened the NIH investigation, and announced that he, too, was calling in forensics experts. It is this second panel whose report is expected soon.

What follows is an explanation of the most conclusive and—given the complexity of the immunological science under investigation—easily understood evidence that *Science* has uncovered.

UNAUTHENTIC DATA

Potentially the most damaging evidence investigators are examining involves allegedly unauthentic data. Even the first investigatory panel, while validating the science and clearing the authors of misconduct, uncovered a number of troubling inconsistencies in the data underlying some parts of the *Cell* paper, especially those data presented in Table 2. This table purported to show that nearly 76% of certain monoclonal cell cultures demonstrated the indirect effect of the transplanted gene, or "transgene." These results seemed to provide striking support for the paper's main thesis that the transgene had influenced antibody production by the mouse's own genes, but the panel noted that the raw data taken from 340 of these antibody-producing cell lines, or hybridomas, seemed in some cases to contradict the results published in the table.

When the first group of NIH investigators asked Imanishi-Kari to account for the discrepancies, she had a simple, if surprising, explanation: The 340 cultures weren't necessarily monoclonal cultures at all—though the paper had stated this fact explicitly—but merely "wells," or cultures that might contain several different strains of antibody-producing cells. The committee was worried by this explanation. As NIH panel chairman Joseph Davie told *Science* in a recent interview: "Unless you have a clonal population where each (recorded) value represents the product of a single cell, it's impossible to calculate (such) antibody frequencies."

The next day, however, Imanishi-Kari reassured the committee by presenting it with several pages of unpublished data from a "subcloning" analysis of these Table 2 wells. Subcloning involves growing, or "cloning," hybridomas from a single cell extracted from a polyclonal culture. The NIH panel was convinced, as member Hugh McDevitt told the paper's authors later at a tape-recorded meeting on 3 May 1989. (*Science* has obtained a partial transcript of that meeting.) McDevitt said that until Imanishi-Kari presented it with the unpublished subcloning data—which confirmed, albeit less strongly, the Table 2 claims—the panel had decided "the whole (study) should be thrown out the window." The subcloning data "convinced us that maybe there was something to the thesis [of the paper]," he continued.

In a recent interview with *Science,* Imanishi-Kari said these critical subcloning experiments were performed on 20–22 June 1985. But Secret Service agents testified in a Dingell hearing on 14 May 1990 that their forensic analysis is at odds with her account. The clearest evidence of discrepancies in Imanishi-Kari's claim presented by the Secret Service comes from an analysis of radiation counter tapes that are fixed to Imanishi-Kari's laboratory pages along with the subcloning data. When biologists want to ascertain the quantity of antibody present in a given solution, they often perform a radioimmunoassay (RIA) in which they "tag" another antibody that specifically recognizes the first with a radioactive label such as iodine-125, then measure the radioactivity with a counter that prints the number of gamma decays on paper tapes. By comparing the paper color, ink composition, and print density on these tapes with those produced by other scientists who used the same counter, forensics experts can date them relatively easily. And since the tapes are produced as an experiment is performed, they should be an accurate indicator of when the work was done.

Such dating is exactly what the Secret Service did at Dingell's behest in the spring of 1990. According to the Secret Service report to the congressional committee, the agents concluded that these particular subcloning tapes "are not consistent with experiments having been performed by other researchers on or around [June 1985]." In testimony at the hearing last May, Secret Service chief document examiner John Hargett explained that Imanishi-Kari's tapes were produced at a time "several months and probably years" earlier or later than Imanishi-Kari had claimed. Because her subcloning tests are one link in a tight chain of experiments recorded in the notebook, a date discrepancy of this magnitude could cast a large portion of the notebook into doubt.

Sources close to the NIH investigation have also revealed that at the request of OSI, Secret Service agents have compared the ink and paper color of these tapes with samples taken from the notes of other scientists who used the same counter in order to date them more precisely. In these analyses, agents reportedly have matched Imanishi-Kari's tapes to tapes in the notebooks of former MIT graduate student Charles Maplethorpe, who performed his experiments in the early 1980s—at least one and possibly several years before the mice used in Imanishi-Kari's experiments had even been delivered to her laboratory.

Imanishi-Kari declined to testify at Dingell's hearing last May, saying later that she didn't know the specific charges brought against her. But she has willingly discussed the matter of the Table 2 subcloning tapes with *Science* and expresses bafflement about the Secret Service results. "Things did happen all the time at those [counters]," she said. "I mean, papers were changed, ribbons were changed . . . There are a thousand and one different [explanations], and I am not the one who can tell exactly why it is different—I have not the faintest idea."

To NIH and congressional investigators, there are still other indications that these subcloning data may not be authentic. For instance, Imanishi-Kari told *Science* that while she pasted some of these tapes into her notebook, she copied radiation count data from others by hand. If the data were generated by a radiation counter, however, digits in the tens columns of Imanishi-Kari's recorded counts (the unit column was rounded off) should be random, yet there is an unusual abundance of 1s and 3s among these digits, and a scarcity of 2s and 9s. When *Science* subjected this distribution to a chi-squared statistical analysis, the result suggested that such a skewed distribution has only one chance in 10^{32} of occurring randomly. Imanishi-Kari admits that she doesn't have a ready explanation for these nonrandom numbers: "I can come up with ad hoc explanations, but I cannot tell if any one is right."

PATTERNS OF CHANGED DATES AND MISORDERED PAGES

Much of the press accounts of the first hearing by Dingell's subcommittee, in May 1989, focused on the testimony of Hargett, the document examiner for the Secret Service. At that time, Hargett had only done a preliminary analysis of the counter tapes, but he had subjected Imanishi-Kari's notebooks to forensic analysis and concluded that they contained at least 25 pages "which raised some question in our minds regarding the authenticity of these pages." Hargett revealed numerous instances where the dates on laboratory pages had been altered and where pages had been backdated, such as a page dated 1984 that was shown through forensic analysis to have been written after a page dated 1986.

But Hargett's testimony raised almost as many questions as it answered. First of all, from the way Hargett presented his findings at the hearing, it was impossible to tell how these alterations affected the scientific conclusions of the paper, if at all. Second, Baltimore, in his prepared testimony before the subcommittee, suggested that the alterations were irrelevant: "[T]he pages which concerned [the Secret Service] contained none of the data that actually contributed to the *Cell* paper." And third, Imanishi-Kari came up with what observers dubbed her "sloppiness defense." She testified that she sometimes didn't date experiments on the days they were performed, that she often recopied old notebook pages, and that she frequently kept counter tapes stuffed in a desk drawer for months before cutting them up and pasting them down on notebook pages.

These Secret Service findings have turned out to be important to the investigation, however. What was not made clear at the 1989 hearing was that some of the data on those notebook pages concerned the subcloning experiments that had played a critical role in convincing NIH's initial investigators to accept the paper's conclusions. Moreover, as OSI deputy director Suzanne Hadley testified a year later before Dingell's committee, in the May 1990 hearing, the Secret Service had questioned the authenticity of some data published in Table 2 that described certain control experiments. As for the sloppiness defense, when Representative Ron Wyden (D-OR) asked Hargett to evaluate it at the 1990 hearing, he testified that "we believe [Imanishi-Kari's] testimony [regarding her explanations] before the subcommittee last year was not accurate."

INCONSISTENT DATA

Worrisome as these assaults on Imanishi-Kari's integrity might be, they do not invalidate all the supporting data for the paper's conclusions. Indeed, to many immunologists, the strongest evidence demonstrating indirect effects of the transgene on the endogenous mouse antibody repertoire presented in the *Cell* paper was not the serological evidence in Table 2 but the molecular analysis of 34 hybridomas listed in Table 3. There, among other findings, the authors reported that several hybridomas produced antibodies with a particular characteristic known as idiotype related to the transgene-even though the cells lacked the transgene itself. So even if Table 2 fell apart completely, Columbia immunologist Alan Stall told *Science,* the evidence presented in Table 3 "is [still] very striking." But the serological data apparently used to demonstrate that the Table 3 hybridomas were producing the transgenic idiotype have also been questioned by congressional and NIH investigators.

A surprising bit of evidence surfaced late last year when these investigators unearthed an NIH grant application submitted on 2 February 1985 by MIT researcher Herman Eisen, Imanishi-Kari, and three other biologists. Imanishi-Kari's data in the application includes a description of the same set of hybridomas from which the Table 3 hybridomas were taken. The characteristics of the hybridomas in the application differ from those in Imanishi-Kari's laboratory notebook, however—and in crucial ways.

For instance, the notebook records that on 12 December 1984 a full 119 out of 147 hybridomas tested positive for the transgene's idiotype, suggesting that the transgene was influencing the mouse's antibody production. But in the grant application, only 60 of 150 are said to have tested positive.

The grant application, whose existence was first reported by *Nature* last September, provides an independent check on the data in the notebooks, and the contradiction would therefore seem to be damaging. But the issue is not clear-cut. Imanishi-Kari told *Science* that the discrepancy resulted from using two different tests for idiotype—a radioimmunoassay in the grant application, and a more sensitive enzyme-linked assay known as an ELISA in the notebook. Immunologists say it is difficult to determine whether this discrepancy could be reasonably attributed to differences in the assays. For instance, the underlying distribution of positives and negatives in the data could have a strong effect on the obtained results, Stall told *Science.*

A direct comparison between the two sets of data might help clear up the reasons for the discrepancy, but Imanishi-Kari told *Science* that she can no longer find the raw data from the RIA. "Very often, at that time, when I made a pile of data, I threw the original data away," she said.

In any case, why didn't Imanishi-Kari submit the ELISA data in the subsequent grant application since they offered stronger support for the notion that the transgene was influencing the mouse's endogenous genes? She told *Science* that the transgenic project was a "minor part" of the grant application, adding that because other tables in the application contained data taken from RIAs, submitting the RIA-generated idiotype data was "a matter of choice—it was the easiest in the context."

In a 10 January 1990 memo to NIH's Office of Scientific Integrity obtained by *Science,* however, O'Toole challenges this explanation. She alleges that Imanishi-Kari did not use an ELISA to test hybridomas for idiotype at all, as the notebook indicates. In her memo, O'Toole claimed that Imanishi-Kari told her in 1986 that she had performed the ELISA recorded in the laboratory notebook to test only for a characteristic known as isotype. O'Toole charged that the reagents used in the isotype assay could not have detected the presence of idiotype.

This dispute essentially boils down to a question of which reagents Imanishi-Kari used in the ELISA and there appears to be no definitive way to check it. Dingell's committee staffers did, however, have the Secret Service examine Imanishi-Kari's notebook pages containing the ELISA data. The first page contains a handwritten statement that an idiotype-detecting reagent was used to screen the hybridomas, but the forensic analysis indicated that this statement was added in a different ink from the rest of the page after the data themselves were recorded.

Such issues have kept members of the current NIH panel occupied for an inordinate amount of time. Their "employer," OSI deputy director Hadley, estimates that the scientists have each put in "hundreds of man-hours." And that time is almost purely advisory: unlike the first NIH investigation, which was conducted entirely by the three immunologists convened by NIH, the new five-member scientific panel defers line duties to OSI staff members. "We do all of the heavy-duty interviewing and data review," Hadley told *Science*. "We do the legwork and present it to the panel. They look, and say, 'You haven't done X, Y, and Z.'"

Will this incredible effort be worth it, if only because it finally puts matters to rest and allows the principals to go on with their lives? Perhaps not. According to Hadley, OSI is already planning a "phase two" of the investigation—dubbed by Dingell aides the "who-knew-what-when" investigation. OSI has passed on responsibility for this follow-up to the inspector general's office within the Department of Health and Human Services.

Whatever the final result of the NIH and other investigations, the Baltimore case has already given some of science's most prominent members and institutions a black eye. And there is little question in the minds of many prominent scientists that the damage has been partly self-inflicted. As Harvard molecular biologist Walter Gilbert, who has watched the case closely, says: "Everyone could have walked away after making a public retraction . . . I'll never know why David [Baltimore] defended the paper down the line like that. There was no reason to defend the paper that way."

Notes

1. D. Weaver, M. Reis, C. Albanese, F. Constantini, D. Baltimore, and T. Imanishi-Kari, "Altered repertoire of endogenous immunoglobulin gene expression in transgenic mice containing a rearranged Mu heavy chain gene," *Cell* 45(1986): 247.

2. The first panel included chairman Joseph Davie, vice president of Searle; Stanford immunologist Hugh McDevitt; and University of Chicago immunologist Ursula Storb. Carnegie-Mellon biologist William McClure and University of Texas biologist Stewart Sell joined the panel when the investigation was reopened.

Misconduct: Caltech's Trial by Fire

LESLIE ROBERTS

Two apparently unrelated cases of alleged scientific fraud in Leroy Hood's huge lab were, by most accounts, handled deftly by Hood and the university.

Caltech, unlike a number of other premier universities, had not been hit with a single case of research fraud—until last year. But when trouble came, it came in spades. Last summer university officials acknowledged that two research fellows in the lab of one of its stars, biologist Leroy Hood, were under investigation for two apparently unrelated cases of fraud. Now those investigations are complete, and both postdocs have been found to have fabricated data—a conclusion that has rocked the prestigious campus. Three papers have been retracted—the most recent just last July. Hood was a coauthor on the papers but was never accused of any wrongdoing.

In stark contrast to the way the principal investigators and their institutions handled the so-called Baltimore case, Hood and Caltech seemed to have dealt with these two cases in an exemplary manner, say Hood's supporters. University officials pulled out their new fraud guidelines, crafted just the year before, immediately launched two extensive investigations, and notified all concerned. Hood swiftly retracted three questionable papers even before the investigations were complete. "That is the right way to do it, instead of waiting and waiting," says James Allison, an immunologist at the University of California, Berkeley—a reference to the Baltimore case, in which a suspect paper was retracted only after several years of wrenching debate, congressional hearings, and Secret Service investigations.

But among all the praise, there is one vocal dissenter: Eli Sercarz, an immunologist and Hood collaborator at the University of California, Los Angeles. Sercarz followed the events closely as they unfolded, and he contends that Caltech acted precipitously in distancing itself from at least one of the accused, denying him due process.

"You're damned if you do, damned if you don't," says a prominent geneticist, who requested anonymity. He notes that David Baltimore, now president of Rockefeller University, has been widely criticized for being too loyal to his colleague, Thereza Imanishi-Kari, while Sercarz is criticizing Hood for the exact opposite.

All of which underscores the fact, he says, that the academic community is still largely working in the dark, without uniform standards on how best to protect the often conflicting interests of everyone concerned. "We don't have rules for behavior in these circumstances," agrees Stanford immunologist Irv Weissman. The Office of Scientific Integrity (OSI) at the National Institutes of Health has general guidelines but leaves it to each institution to craft its own procedures—none of which can possibly anticipate every quirk and twist likely to arise. Faced with myriad judgment calls along the way, university administrators are essentially winging it, learning as they go. And for Caltech, it was trial by fire.

(The two postdocs accused of fraud declined repeated requests for interviews, though one of them, Vipin Kumar, provided a short written statement. This account is based on interviews with several people involved or close to the investigations, some of whom requested anonymity, and two written statements from Caltech.)

DOCTORED FIGURE PROMPTS TWO PROBES

Vipin Kumar and James Urban joined the Hood lab several years ago: Kumar from a postdoc at Harvard, Urban from the University of Chicago. They began working, at first together but then independently, in an especially hot area of immunology research, looking at the molecular biology of and possible treatments for autoimmune diseases such as multiple sclerosis. Pressure was intense, as it is throughout the huge Hood lab, which numbers 65—especially because Hugh McDevitt's group at Stanford was pursuing the same tack. Both Kumar and Urban were ambitious, logging long hours and winning high marks from Hood in the process.

But not everyone shared Hood's opinion—and several people in the group went to him with their suspicions, not about fraud, per se, but about sloppy science, says Hood. He investigated each accusation and turned up nothing solid, chalking the problems up to personality conflicts and inexperience. "I had complete faith," he recalls. Indeed, Hood would be the last one to suspect fraud, one source said, alluding to both his honesty and perhaps, his naivete. "Lee doesn't like to believe things like that. It is the last thing he would expect someone to do."

That faith began to crumble in late May of 1990, when Dennis Zaller, a senior member of Hood's group who is now at Merck, Sharpe, & Dohme Research Laboratories, and a colleague went to Hood with what they thought was clear evidence of wrongdoing. Zaller had been trying to extend some of Kumar's work, and in the process tried to repeat one of his experiments. He couldn't. He then showed Kumar's original paper, which had been published in the December 1989 *Journal of Experimental Medicine* (*JEM*), to Mike Nishimura of the Hood group. Nishimura was struck by what everyone in the lab, including Hood, and the *JEM* peer reviewers had missed the first time around: a key figure appeared to be falsified.

Says Zaller: "If you look at the [Southern] blot it is unmistakable." It was supposed to show DNA from several different cell lines that all had essentially the same pattern—namely, a rearrangement in the T cell receptor gene locus. But Zaller and Nishimura could tell by looking at the artifacts, the little spots that crop up on gels, that Kumar had used data from just a few cell lines—one lane in each—duplicated repeatedly and labeled as if they came from many more cell lines.

A stunned Hood immediately informed the chairman of the biology division and other university officials, who began an inquiry into the allegations—the first step to see whether a full investigation is warranted. While the inquiry was getting under way, Hood enlisted the senior scientists in his group to perform an internal review of all of Kumar's work; Hood later gave their report to the investigation committee. He also asked others in the lab to try to repeat the *JEM* experiment. They couldn't.

But that wasn't the only devastating finding. In the process of reviewing Kumar's data, the Hood group looked into some of Urban's work as well, as he was a coauthor on some of Kumar's papers. To their dismay, they quickly spotted what looked like a problem in his work, too—a problem that appeared to be unrelated to Kumar's alleged misdeed. Hood found himself in the unenviable position of telling university officials that his lab might have a second case of misconduct on its hands. Caltech vice president and provost Paul Jennings launched a separate inquiry, which got under way on 20 August 1990.

THE KUMAR INVESTIGATION

When Hood confronted Kumar, asking him to provide the original data and explain how he had constructed the Southern blot, Kumar reportedly did not deny doctoring the figure but did deny any intentional fraud. Instead, he insisted that he had only been trying to create a more attractive image and that he did not know this sort of duplication was unacceptable practice, explains UCLA's Sercarz, his staunch defender. Indeed, Sercarz says Kumar sought advice from Urban, his "mentor" in the lab, on the propriety of duplicating lanes but apparently misunderstood what Urban told him. Says one Caltech source: "His rationale was essentially, 'I was young and naive.'"

Sercarz, for one, buys that argument, explaining that "Vipin had never prepared a paper before." In India, where he studied at the Institute of Science in Bangalore, his adviser wrote most of his thesis, says Sercarz. And when Kumar went to Harvard for his first postdoc, says Sercarz, his professor, Debajit Biswas, prepared all the papers and figures—a fact Biswas confirms. Says Sercarz: "Vipin arrived at Caltech a very, very, green fellow. Vipin did not know what to do with lanes that were irregular. He wanted to rationalize it to produce an esthetic figure." Sercarz notes that Kumar made no effort to hide the telltale artifacts. In fact, he used the Caltech photographer to prepare the figure. "The behavior of someone deliberately falsifying something is different than that," he contends.

But the inquiry committee, which met with Kumar, did not buy that defense. "If that argument had carried the day, the outcome of the investigation would have been very different," says Jennings. The committee, chaired by the head of the biology division, decided just one week later, on 8 June, that a full-blown investigation was warranted. Jennings set up a committee of four members of the biology division to investigate. It began working on 13 June. During the investigation, Kumar was relieved of his duties in Hood's lab, though he retained his appointment there. Says Jennings: "We did not want to act until the investigation was complete." About that time, Hood and Jennings decided that, rather than wait for the results of the investigation, Hood should retract the *JEM* paper, since Kumar had admitted duplicating the lanes, though he denied fraud.

The investigation had come at an extremely awkward time for both Kumar and Caltech. Nearing the end of his postdoc, Kumar had applied for several jobs, with strong recommendations from Hood. After considerable soul-searching, Hood and Jennings decided they had no choice but to notify the universities to which Kumar had applied, along with the journals that had published the suspect work, coauthors, the National Multiple Sclerosis Society, which had given him a fellowship, and, as required, NIH, which had funded the work, and the National Science Foundation, which supports Hood. "We tried to do it as confidentially as possible," says Jennings, but before long the community was abuzz. Washington University in St. Louis, which had already offered Kumar a job, withdrew its offer.

Kumar took the developments hard, having what Sercarz and others describe as a nervous breakdown that required hospitalization for several days. After that, Sercarz, who knew Kumar well through his collaborative work with the Hood group, took Kumar into his lab to continue his research while the investigation proceeded, though he officially retained his position at Caltech.

It was Hood's letter to immunologists at the universities where Kumar had applied that Sercarz feels was inappropriate. Argues Sercarz: "It is a precipitous action to deny due process before there is an investigation. They took away this man's livelihood—how can that be fair? The matter was spread throughout the country before there was a real investigation. It was unfair to tarnish his reputation."

Responds Hood: "It was a complicated call, how much to get other people involved. We talked a lot about it at Caltech. Everyone who could have been affected should have been notified. Fraud can't be brushed under the rug. If he had been cleared, I would have written a letter to everyone explaining what had happened."

Jennings, too, defends the letter. "Our rationale was that we had general responsibility to the scientific community for the stewardship of scientific research. I still think it was the appropriate thing to do. The letter went out on a need-to-know basis. We presumed that people would act fairly and wait to see how the investigation came out."

THE KUMAR FINDINGS

The Kumar investigation was completed in late March. Caltech officials will not release the reports on either of the investigations to the public or the press. They did, however, release two short statements to *Science* describing the resolution of both cases. The statement on Kumar mentions only the fabricated Southern blot in the *JEM* paper, which the investigation committee decided was research fraud—the most serious offense in Caltech's judgment, and one that implies intent to deceive, explains Jennings. But one source talks of the "pervasiveness of the problems" uncovered. "There was a lot of work that was not well done; a few too many corners were cut in an unethical fashion."

The investigation turned up problems with another altered Southern blot in a paper submitted to *Cell;* that paper was withdrawn before publication. Perhaps the biggest problem was that much of Kumar's data was missing, say several sources. Kumar maintains that two of his lab notebooks, containing his scintillation counter data, were stolen—in fact, he had informed Hood about the theft before the fraud accusations arose. Even so, says Hood, Kumar never informed him that all of his data were missing for one paper, a paper on autoimmunity published in the February 1990 *Proceedings of the National Academy of Sciences.* Hood retracted the paper this July. "I think most things in those [retracted] papers were correct," says Hood, "but because of what had happened, I could not leave the papers out there with no data to back them up." Speaking through Sercarz, Kumar denies that most of his data were gone, insisting that all the data were available for the *JEM* paper and that some data were available for other papers.

Kumar's Caltech appointment was terminated on 31 March. To this day Kumar, who is now a postdoc in Sercarz's UCLA lab, contends that "I am not guilty of any deliberate wrongdoing," as he wrote in a brief statement to *Science,* without elaborating. Indeed, when he learned of the resolution he appealed to Caltech's president, Thomas Everhart. After reviewing the file and meeting with Kumar, Everhart decided against reopening the investigation, says Jennings.

THE URBAN INVESTIGATION

The Urban case was more complicated, almost from the outset, say Caltech sources. Hood's group first turned up what looked like a fairly minor problem with a figure, which prompted the inquiry, says Hood. But the investigators soon encountered more serious problems as well. Caltech decided to extend the inquiry to give Urban time to respond to all the charges. Thus, the full investigation did not get under way until 16 November, with a committee of five faculty members, four from inside the department and one from without. Those familiar with the case say the committee found no sign of collusion between the two postdocs but multiple problems in Urban's work, as in Kumar's.

The problems with Urban proved trickier to nail down initially. "The smoking gun did not exist in the same way," says one source, referring to Kumar's Southern blot. As in Kumar's case, much of Urban's data was missing—discarded, Urban told the committee, when he moved to the University of Chicago in March. What the investigators did find was that the final version of a paper exploring a mouse model for multiple sclerosis, published in the 20 October 1989 issue of *Cell,* contained different data from the one that had been submitted for review. The committee concluded that the data in the first draft were fabricated—a charge Urban reportedly did not deny, though he did deny any intent to deceive. According to one official close to the case, Urban said he intended to do the work and assumed he knew how it would turn out. "All along, he claimed he was just trying, because of pressure, to speed the review process and that he never intended to publish without real data." No one in the lab, including Hood, noticed the discrepancy between the review draft and the final version; nor did *Cell.* Unswayed by Urban's explanation, the Caltech committee decided that his actions constituted serious scientific misconduct, a slightly lesser offense than fraud as it does not imply intentional deceit, explains Jennings. The *Cell* paper was subsequently retracted.

The Caltech statement released to *Science* also alludes to other instances of poor judgment or misconduct. The problem, explains Zaller, who was involved in Hood's internal lab review before moving to Merck, is that, "We don't know what he did or didn't do. We have no way to evaluate how much he actually did because some of his lab notebooks were missing." They also confirmed problems with the figure that had prompted the inquiry. Says one source: "It was quasi-made up but more or less reflected what the real data were. It was not an obvious attempt to make up data."

According to one source, Urban accepted responsibility for his actions—essentially admitting that he had done bad science—though he did deny intentional fraud. His only excuse was pressure, says that source, "but he did not push it. In the end, he seemed relieved." Because Urban had already left Caltech, the only possible sanction was a letter of reprimand, which was sent on 24 June. Urban had informed Chicago officials about the investigation, and at their request, Caltech informed them of the resolution as well. Urban, a pathologist by training, resigned from the University of Chicago immediately thereafter.

The Caltech reports have now gone to OSI at NIH, which is reviewing them to see whether further investigations are warranted. If it accepts Caltech's findings, the Public Health Service will in all likelihood impose additional sanctions, says Alan Price of the fraud office. These could range from banning Kumar or Urban from serving on Public Health Service advisory committees to debarring them from receiving federal funds for perhaps three to five years.

"It's a sad day," said Stanford's Weissman, when informed of the final verdict for both Kumar and Urban. "Jim Urban applied for a job as a postdoc. On paper and in person, he was extremely positive. I offered him a place in my lab but he didn't come. He was very, very promising." Another talks of "how sad it is to watch two careers crumble."

LINGERING CONCERNS

The only real gripes that *Science* uncovered about Caltech's handling of the two cases—or that Caltech officials have heard, other than Kumar's objections—come from Sercarz, who feels that Caltech's procedures, like OSI's, deny the accused due process. Aside from criticizing Hood for the letter he sent to everyone Kumar had approached for a job while a postdoc in Hood's lab, Sercarz faults Caltech for the length of the investigations—which both took about a year—a shortcoming Jennings and Hood concede. And he argues that Caltech should have allowed the accused and his accusers to meet face to face and to cross-examine each other. Sercarz and others have leveled similar complaints at OSI's own procedures (see *Science*, 6 September, p. 1084). Kumar, too, says in his statement to *Science*, "during the investigation procedure, there were many violations of due process by Caltech."

On a more fundamental level, Sercarz also questions whether a university with an interest in protecting its reputation can really be impartial. "No one knows what the ideal procedure is. But when the principal investigator [lab chief] is someone powerful like Lee Hood, the university may want to decrease his involvement in the alleged misconduct and blame everything on the postdoc. That could lead to a distortion. In general, having an external committee of experts might make the investigation more impartial."

It is now up to OSI to determine whether Kumar got a fair shake or whether, as Sercarz believes, there are lingering questions.

THE AFTERMATH

The Hood group is now recovering from what has been a very tough year. Says Hood: "It was a traumatic experience for everyone involved, not just for the accused but for all around them." Like everyone else, they are wondering how it could have happened—and how to prevent it from happening again.

Hood and his co-workers are now trying to replicate some of the crucial experiments performed by Urban and Kumar. Says Hood: "We can't redo it all. It is a tremendous amount of work." He has also instituted tighter controls in his lab. The committees didn't find any "major shortcomings" in Hood's procedures, says Jennings—in fact, Jennings calls them "pretty good"—but there was obviously room for improvement. "You would hope the procedures would pick up the problem," says Jennings. Hood has now formalized the review process, so that each paper is now reviewed by three people inside the lab. There is considerably more emphasis on dealing with raw data, not merely a synopsis of findings. And Hood now also requires everyone to keep a bound lab notebook—and has made clear that it is the property of Caltech, not of the scientist.

When the dust settles, Caltech officials plan to take a look at how well they handled their trial by fire, to see if any of their investigatory procedures should be changed. In the interim, faculty members are debating whether to offer a course for new graduate students on the rules of scientific conduct. Explains Jennings: "The community has always figured that you just know how to do these things, such as how to handle data. But maybe people would benefit from a course spelling out the rules on keeping research data. It would be an opportunity to ensure more formal acquaintance with issues and procedures we used to take for granted."

4

Relationships In Laboratories and Research Communities

VIVIAN WEIL AND ROBERT ARZBAECHER

I. INTRODUCTION

In November, 1993 the National Institutes of Health (NIH) dropped charges of scientific misconduct against AIDS researcher Robert Gallo, thereby ending a four-year investigation. This action came a few days after the Health and Human Services Departmental Appeals Board rejected charges against his former associate, Mikulas Popovic, and a few days before the Appeals Board was to hear Gallo's arguments. NIH's reversal did not provide closure. According to one observer "Like an underground fire that won't go out, resentment continues to seethe over the official exoneration of the renowned Gallo."[1] The case had begun with the charge that Gallo had misappropriated from a French research group the AIDs virus he is credited with co-discovering. After initially claiming that AIDs was caused by a virus he had discovered, Gallo conceded that in the discovery process there might have been accidental contamination from the virus supplied to him by the Pasteur Institute in France.

Controversy continues about whether Popovic, Gallo's subordinate in the laboratory, deliberately used the French virus to grow the cultures for identifying the AIDs virus and whether, if he did, Gallo knew it. Controversy also surrounds a report of their work on the AIDS virus that Gallo and Popovic published in which references to the French virus were deleted. In a report to NIH's director evaluating OSI's findings, Dr. Frederick M. Richards (a chemist from Yale who was a scientific consultant in the investigation) charged that Gallo's lab deliberately hid its dependence on the French virus. He also criticized findings of NIH's Office of Scientific Research Integrity for placing blame on the subordinate and for failing to blame the lab chief, "who had the duty to 'monitor the performance of all personnel in the Laboratory and to pay particular attention to the accuracy of major publications which bear his name as author.'"[2] The case was reopened, with the result, a year later, that NIH issued two findings of misconduct against Gallo. Those were the findings that NIH dropped in anticipation of the Appeals Board hearings.

The relevance of the Gallo-Popovic story for our purposes is that it draws attention to the activities within research groups. It brings to the foreground features of the daily conduct of research that matter a great deal to members of research groups as they go about their work; that are significant from the perspective of research ethics; and that do not often become public. This bench-level view of scientific research throws light on relationships between the laboratory director and subordinates, the preparation of reports of research for publication, the standards for and responsibilities of authorship, the basis for giving credit and recognition, and the handling of data or materials from other research groups.

The ethical significance of the management of research groups may not be evident from instances of clear-cut misconduct. The inclination is to describe misconduct as the misdeeds of isolated individuals, "bad apples" somehow not weeded out in time. In contrast, instances such as the Gallo case leave insistent questions about the character of the research environment and the management of the research. The cases force observers to consider the role that the management of the research group plays in giving rise to distrust, allegations, and unethical conduct. The cases show the damage that can result from suspicions, anger, an air of intrigue, and allegations. If the expense of investigating were the only cost, the damage would be significant. But injuries to individual reputations, careers, to areas of research, and to institutions add incalculable harm. This is not to mention emotional damage, harm to human relations and friendships, and destruction of careers.

The way a research group is managed has to be the entire group's business. All research groups have an interest in fostering a climate of trust, an atmosphere that supports "responsible" conduct, conduct that meets justified standards of the scientific community. This sense of "responsible" takes in more than the avoidance of falsification, fabrication, and plagiarism, the usual examples of misconduct. It includes some ethical standards that are presumed to apply across research communities, as well as standards specific to the local research community. Each research group should make explicit the underlying assumptions and rationales for their standards, while recognizing that standards evolve over time, as circumstances change and new problems come to light. Ensuring that new researchers who join the group, including graduate students and postdocs, understand the prevailing standards is part of the group's business.

Most of the situations examined in this chapter do not rise to the level of public scandals; they are not the kinds of cases that spill out into the pages of Science or Nature. Rather, they are of the sort that are born in the lab or research group and usually die there. Nevertheless, they involve practices that are critical to the progress of research. When an advisor fails to acknowledge a graduate student's refinement of a technique in a paper utilizing that refinement, the student may express his or her consternation to no one beyond fellow students or friends. In the absence of guidelines or standards regarding recognition, he or she may be unsure about how to react. Yet giving credit for research and taking responsibility for reported research are core concerns of the research enterprise. Failure to articulate standards with respect to these core concerns can significantly damage the research process and its products, the researchers in these situations, and the research environment.

II. FEATURES OF THE RESEARCH ENVIRONMENT

We often hear that a set of shared values including "honesty, integrity, objectivity, and collegiality" binds scientists into a wide, international community.[3] But the research enterprise is highly decentralized. In the university and company departments and laboratories in which scientists conduct their investigations, they operate under local assumptions about how research should be managed and how members of research groups should relate to one another. New apprentices, graduate students and postdocs, learn how to conduct research from local practices they encounter. Local control allows for "diversity, flexibility, and creativity."[4] However, the relative autonomy of research groups that permits them to establish their own procedures and standards, imposes on them the burden of figuring out how to incorporate any shared values and live up to accepted standards.

Personal characteristics and styles of behavior, especially of research directors, significantly affect the research atmosphere. And research groups are vulnerable to the kinds of misunderstanding and conflicts (e.g., rifts arising from jealousy) that can mar any cooperative enterprise. Moreover, in the view of many, the competitive environment of science in the United States imposes additional stresses on the human relationships and practices within research groups. Comments from insiders occasionally provide glimpses of how life in the competitive atmosphere of research groups can fail to mirror the expected values. Barbara Kingsolver, a former graduate student in the field of population biology and now the author of best-selling fiction, recently explained that she gave up graduate study after "growing tired of the grinding lab work . . . and tired of the academic back stabbing."[5] An official in a position to hear allegations of misconduct remarked recently that there are "unhappy" labs out there.[6]

These hints of the potential for the breakdown of trust underscore the need to look more closely at relationships in research and laboratory groups and at the ways in which research groups are managed. A research environment where relationships are distant, frayed or fractured—an "unhappy lab"—may well not sustain responsible research conduct. In such environments, distrust can grow and lead people to entertain suspicions or to engage in countermeasures that violate standards. Trust is of key importance to the enterprise of science.

There is diversity in the way research groups are organized and managed and in the methodology of investigation. Though there may be some common methodologies, methodology itself can be the substance of scientific debate. Contrary to popular notions, we do not find an abstract, universal " scientific method" that guides practice in all situations, whether we are talking about how research is managed in research groups or investigative methodology.

Research groups and laboratories are differentiated by scientific discipline, by fields and research programs within a discipline, by host institution, and by department within the institution, as well as by the personality of the director of the research group. We cannot expect to capture here this great variety of relationships and practices at the bench level. Furthermore, we assume that a wide range of morally permitted practices and ways of managing research projects and human relationships can provide conditions for responsible conduct. We concentrate on university research settings because of their centrality in training scientists and also because of their accessibility. By emphasizing on what is at stake in the practice (of assigning credit, for example), we expect to give guidance without specifying what local arrangements should be adopted.[7]

Some amplification of what the terms "bench-level science" and "individual laboratory and research groups" refer to should make clear what is under scrutiny. Many research groups are based in a lab, but that is not a requirement for inclusion. Groups of up to 20 people that consist of a principal investigator or project director, an additional investigator or more (who may include peers, junior investigators, postdocs, or graduate students), with at least one trainee, will be encompassed. In agreement with many observers, we regard the principal investigator or director of a group of more than 20 as functioning more like a business manager than a manager of science. Certain risks are associated with managing research on such a large scale. Occupied by administrative tasks and the pursuit of funding, the manager may not be sufficiently accessible, may lack hands-on familiarity with the conduct of research, and may fail, if only by mere absence, to provide a model of appropriate standards. Many observers agree that such conditions are not conducive to a research group's articulating and maintaining standards of responsible conduct.[8] In these very large groups, observers believe, appropriate ethical standards are less likely to be transmitted to trainees.

Three major goals characterize and help to identify research groups: (1) to get research done; (2) to get students trained; and (3) to acquire the funding needed to achieve the first two goals. The first two are generally regarded as intrinsically valuable goals. The third goal is instrumentally valuable for achieving the first two goals. Whether explicitly announced or not, these goals drive the group's activities and command the director's attention. The emphasis within this mix of objectives varies from research group to research group and helps to create the characteristic "flavor" of each group. Some make training more central, some research; and some pursue funding with special ardor.

Research groups are cooperative enterprises. In order to accomplish their goals, they depend on members each doing their part. And members depend upon one another. Participants in a cooperative enterprise must be able to accommodate their personal concerns to the objectives of the group. Sometimes observers write as if the cooperative, collaborative features were all that characterized these groups. In addition, however, research groups value independence in each of the members; it is a trait that the training aims to foster in students.

Moreover, competition pervades, in the broader structures and systems of science and within research groups. The competition among research groups for research support from funding organizations is a driving force. Investigators compete to achieve recognition through publication. Graduate students and postdocs compete for admittance to research groups. Those groups compete for promising trainees. Researchers, including graduate students and postdocs, compete within the group for recognition of their findings. In the view of some philosophers and historians of science, competition for recognition has fueled the scientific enterprise since the seventeenth century and continues to do so.[9] Moreover, many who are influential in science and in the funding of science in the United States believe that competitive features that have been deliberately incorporated into the conduct of research in this country are responsible for the strength of American science.[10]

Science at the bench level is also characterized by disparities of power. By winning funding, usually external funding, the principal investigator or project director acquires the resources for research; control of those resources constitutes the basis of that individual's power. Other members of the group, especially any junior investigators, postdocs, and graduate students, are dependent on the director for resources and for career advancement. For the way they wield power within the research group and manage relationships with these dependents, research group directors are relatively free of accountability.

Postdocs are particularly vulnerable since their connection with a single faculty member is usually the basis for their status in the research group, and postdocs do not have their own standing in the university as graduate students do. Some observers liken the postdoc's position to that of a contractor. Though postdocs may make important contributions such as bringing a new technique to the lab or giving guidance to graduate students, generally speaking, they have little leverage of their own.[11]

There are, of course, legal restrictions, government funding agency requirements, and university rules that circumscribe the power of research directors and senior investigators. Nevertheless, a very significant power differential remains and its management is a matter for local determination.

III. OVERVIEW OF ETHICS ISSUES IN RESEARCH ENVIRONMENTS

A. Setting Standards

All research groups require some set of standards or ground rules for the way they operate in proposing, conducting, and reporting on research. These standards or rules may be explicit, even expressed as formal rules, but often they are taken for granted rather than spelled out. One laboratory director may announce, "There are two rules regarding openness in my lab: (1) No one should keep data to himself in the lab, and (2) Within the lab, everyone's opinion has to be open to criticism and discussion." Another might say, "In my lab we take those rules for granted without ever announcing them." In yet another lab, these will not even be recognizable by the lab director as governing norms or rules.

Ordinary morality consists of "those standards everyone wants everyone else to follow even if everyone else's following them means having to follow them oneself."[12] Here, we are concerned with the special standards of conduct everyone wants everyone else to follow, in scientific research communities.[13] Within the research group, this requires the establishment of clear and consistent ground rules or standards. These can range from informal policies to highly codified statements. A question to address is how well the standards cover the range of situations that arise in research groups and require consistent responses. Further questions concern whether the standards reflect the proclaimed values of the scientific enterprise and whether they are acceptable from the wider perspective of ordinary morality.

The professional associations of some scientific disciplines, such as the American Chemical Society and the American Physical Society, have adopted codes of ethics. In this respect, these scientific societies are like the associations of major professions, such as law, medicine, and engineering. It is reasonable to regard the standards and practices that research groups adopt to assure responsible conduct as species of professional standards. The process of articulating standards in research groups can be read as marking a transition from a stage in which unstated, informal standards presumably operated, to one in which the more formal standards of a profession are called for.

B. Cooperation and Competition

That scientific projects require collaboration in an atmosphere of competition has already been emphasized. Research groups have leeway in determining the conditions for collaboration. This is shown by the way research communities vary in the extent to which they value and encourage open communication, an underpinning of cooperation. Local choices also shape competitive arrangements and practices within the research group. There are, for example, project directors who ask more than one student or postdoc to work on the same research problem, more as competitors than collaborators. This practice is controversial, especially where competition within a research group mirrors the "winner take all" feature of competition in the organization of science outside the research group.[14]

The centrality of competition in science may seem a "natural" feature of science as a human activity, especially to those already used to competition in our system of education. Hence, the degree to which competition results from conscious decisions may escape notice. Certain forms of competition flow from certain policies, for instance, policies governing the funding of science. Government grants for essentially the same research projects are awarded to more than one investigator, to many if the research area is important.[15] While we have grown used to the system, we could have a competitive funding system that does not operate this way.

Ethical questions concern the terms of collaboration: How open is information within the group and how easily shared? Is credit determined by clear criteria that apply to everyone? What are the expectations of reciprocity, loyalty, and collegiality? These are relationships characteristic of research groups; collegiality is a relationship specific to professionals. Though generally positive, these relationships can generate expectations of conduct that violates ethical standards. Such a situation arises when "loyalty" could mean ignoring conclusive evidence that a recently appointed senior colleague in one's research group plagiarized the work of a junior investigator at his previous institution. Ethics sets limits on loyalty and collegiality.

Likewise, ethical questions arise concerning competitive arrangements and practices. The competitive atmosphere outside the research group can be mirrored within the group. Recent research shows a connection between a competitive climate within a research group and increased likelihood that a student will observe misconduct over time.[16] The research suggests that a highly competitive atmosphere within a research group can be linked with erosion of trust within the group. There is therefore an ethical issue in determining what forms of competition internal to the research group, if any, are acceptable.

It is not hard to see how demands in the name of competition can come into conflict with demands in the name of collaboration. A postdoc may be reluctant to share recognition with a graduate student with whom she collaborates when she has made her own interesting findings in connection with the graduate student's project. While conflict is not inevitable, there seems to be an inherent tension that arises from the value placed on both collaboration and competition.

C. Power Disparity

Disparities of power in relationships and transactions pose risks of exploitation or abuse of those with little power and frequently precipitate questions about fair treatment. In scientific research, this asymmetry of power is located within a web of relationships that includes those between supervisors and their graduate students and postdocs, between postdocs and graduate students, among graduate students, between senior researchers who are peers and between junior researchers who are peers, and between senior and junior researchers. Technicians also can be important participants in this web of relationships. The power differential causes concern because the parties with less bargaining power or information may consent to arrangements that, with more information or a stronger bargaining position, they would be rational to reject. It can be ethically wrong to impose such arrangements even though the less powerful agree to them.[17]

Among graduate students and postdocs, it is not unusual to hear the term "slave" for graduate student and "indentured servant" for postdoc. There are abundant opportunities for taking unfair advantage that well-meaning people may not notice because of their subtlety or because they occur in ways long taken for granted. In view of expectations of collegial support, even senior members of a group who notice abuses of power may have difficulty doing anything about them. The fear of jeopardizing career prospects makes it especially difficult for junior people to call attention to perceived abuse of power.

Even where project directors are sensitive to the vulnerability of the less powerful members of the group, they may fail to identify situations in which weaker members are taken advantage of and are afraid to speak up. Assigning heavy teaching responsibilities to a postdoc, for example, relieves the postdoc's advisor and other senior researchers of certain burdens, but hinders the progress of the postdoc. At field sites outside the university, the risks of abuse of power or of inadequate supervision are increased. Similarly, at these locations, failures by advisors or trainees to meet standards can have amplified impact.

D. Mentors

Because much of science teaching is one-on-one, the relationship between mentor and student becomes critically important in educating students and transmitting ethical standards. A mentor is not necessarily the project director or the student's thesis advisor. Rather, the mentor is a more senior, experienced person in the research setting who over time provides the student knowledge and nurture with respect to technical matters, professional values, ethical standards, and with respect to building a career in science. Mentoring is identified with taking a sustained, active part in fostering the careers of postdocs and graduate students, not merely with functioning in the role of advisor or instructor.

The term "mentoring" refers to an interactive process; the role of the mentored person is not a passive one. That person has a responsibility to seek information and guidance and to be ready to make use of it.[18]

Graduate students and postdocs benefit from having multiple mentoring relationships, some probably short-term and others of longer duration. Advanced postdocs and graduate students can mentor more junior colleagues. Some recent research suggests that the quantity of mentors may be even more important than the quality of specific mentoring relationships.[19] In any case, whether there is adequate mentoring of all trainees is a major ethical concern for a research group. Because mentoring entails a personal relationship involving a commitment to a trainee above and beyond what is required by the role of advisor or teacher, it is difficult to assure that all graduate students and postdocs are adequately mentored. The success of the mentoring relationship depends upon particular personal qualities of the parties and their degree of commitment. The mentor role, accordingly, is an informal role and cannot be mandated. Recognizing that individual mentoring cannot be assured, some educators recommend "bulk" or "wholesale" mentoring. They have in mind programs that at least support graduate students asking the questions they need answered about the terms of their training and their progress. Such programs may include workshops, short courses, and seminars that transmit ethical standards and help graduate students develop professional skills.[20]

Relationships within research groups can go sour in a great number of ways, including when lines of supervision are not clear; when research problems are not clearly demarcated and allocated; and when well-established lines of and regular occasions for communication are lacking. We can collect these ways of going astray under broader ethical questions about how to wield power responsibly and how to behave responsibly as one dependent on the power of others. As we proceed to point out the kinds of standards and practices that are needed, we thereby delineate role responsibilities in research groups. To fail to fulfill these role responsibilities would be to behave irresponsibly, that is, unethically.

IV. ANALYSIS OF VIGNETTES

The vignettes that follow highlight aspects of the organization and management of research groups and focus on certain relationships within them. The commentary on the vignettes is guided by the concern to identify ethically defensible, responsible courses of action. The analysis also highlights practices and procedures that labs and research groups might adopt to support ethically responsible conduct or reduce the likelihood of such problems occurring.

Vignette 1: The Lab of Last Resorts[21]

The following memo is an example of what a laboratory director might assume to be his or her responsibilities to, and expectations of, the graduate students in the laboratory. By stating the terms of the agreement explicitly, the director hopes to avoid later misunderstandings.

To: New & Used Graduate Students in the Laboratory of Last Resorts
From: Director Drake
Subject: General Rules

Welcome to our laboratory. As you know, research in this laboratory is funded by grants from NIH, NSF, and other agencies. The projects so funded have specific aims and a detailed research plan stated in the grant applications. Departure from these aims and plans requires re-application for the grant funds. We would only do this if the original ideas prove early to be without merit.

Therefore, students in the laboratory are not free to pursue ideas and activities of their own design, unless these fit the aims and research plan of the project that supports them. In accepting this fact you are surrendering a significant amount of intellectual freedom. It is important that you understand what you will gain here and what you will give up. Please be certain that the mutual agreement stated below is acceptable to you.

I agree to provide, as long as grant funds are available:

1. Your tuition.
2. A stipend to live on.
3. Excellent laboratory facilities, including all necessary computers, instruments, equipment, tools, supplies, and desk space.
4. Superior research training.
5. Thesis idea and guidance.
6. A long-term commitment to your career goals.

You agree that, since the Laboratory's highest priority is continued funding, I may:

1. Set your daily work schedule.
2. Determine your research.
3. Personally present your work whenever and wherever I deem it appropriate.
4. Decide what and when to publish.
5. Decide the authorship and order of names on all publications.
6. Determine your readiness for PhD qualifying, preliminary, and final examinations.
7. Approve your committee membership.
8. Approve any communication you have with other laboratories.
9. Have exclusive ownership of your data—before and after you leave the laboratory.

10. Restrict your lunches to the usual Banana and an occasional Tuna Sandwich.

Of course, those who produced this statement are not entirely straightfaced; they are somewhat self-mocking. Other scientists have commented, however, on how closely this statement corresponds to research life in many groups. In any case, the very fact that these ground rules are written down counts ethically to the credit of this research group. That is because everyone is on notice about certain important expectations. It would be even more to a research group's credit if it made such a statement available to prospective students before they accepted offers of admission. Prospective students would be less at the mercy of unrealistic expectations about what lies in store. They might even be in a position to compare different options for graduate study with respect to the stated terms under which graduate training is managed, as well as with respect to the intellectual opportunities. Once a graduate student has signed on to a research group, it is usually difficult to switch to another group. Advance familiarity with the ground rules would help to prevent students from becoming trapped and feeling they were "lured by false advertising."

Some see this research group director as excessively controlling and paternalistic, unwilling to allow students enough freedom to learn to make choices for themselves in order to achieve their educational goals. If the terms for students are too restrictive, it remains to consider what is a fairer bargain, what are the standards of fairness, and how to draw the lines circumscribing students' freedom. There is plenty of room for alternative lists of ground rules.

Often, the basis for control over a student's time and efforts in the lab is the student's status as employee. When that is the case, then the control should be restricted to hours in which the student is paid to perform as an employee. "Every Waking Hour," the half-humorous title that one research group put on its statement of ground rules, expresses an excessive demand, if the student's employee status is what justifies controlling the student's activities. Admittedly, there is often a blend of work and research, but that does not justify nearly total control over the student's time in the lab. This issue brings out complexity in the role of the graduate student. Some graduate students receive support from funds in grants paid to research assistants (RAs). For some, support consists of stipends from traineeships. Some graduate students are hired as teaching assistants (TAs) rather than RAs; the TA position, with its teaching responsibilities, generally poses more problems than the RA position for getting thesis research done. Depending on status, length of time in the program, or whether one takes the perspective of the lab director, advisor, or the person herself or himself, the person can be described as student, junior colleague, employee, slave, or some mixture of these.

Determining and assessing the rationale for each of the prerogatives claimed by Director Drake of the Lab of Last Resorts would help to decide if they are overly restrictive. It is worth noting, however, that many students in humanities disciplines where they are left largely to their own devices might be willing to exchange some freedom for some of the paternalistic support offered by the Lab of Last Resorts.

It has to be acknowledged that making explicit what was formerly implicit can change the atmosphere. Some will regret the loss of spontaneity and informality. And some practices formerly taken for granted will not survive when they are examined and discussed. However, the link between clear expectations and responsible conduct is so strong that it justifies the risks associated with formulating explicit ground rules. Two caveats should be kept in mind. First, the ground rules have to be general; many details of life in research groups must be managed by extrapolating from the ground rules. Second, the ground rules are not without exceptions. For example, the project director may have an explicit policy of allowing postdocs to take all their data with them when they depart, but with the understanding that on occasion that policy may have to be overridden.

The issue of exercise of power is linked with the issue of "ownership" of ideas in the following situation that concerns a graduate student. Focusing on the advisor/graduate student relationship at an early stage, this vignette continues exploration of the vulnerable situation of graduate students.

Vignette 2: Whose Thesis Problem?[22]

George Alvarez, a graduate student and a lively, imaginative fellow, comes up with an idea that is worthy of a thesis and appropriately related to a funded project of the laboratory. However, since Alvarez is only a third of the way through the first year of the program, the Lab Director, Helen Jonas, judges that he is not yet ready to pursue the idea. He has not acquired the background and skills to carry through the research. Jonas therefore assigns Alvarez's idea as a thesis problem to another student, Charlene Wright, who is ready to pursue it. Jonas assures Alvarez that there will be an interesting problem for him when he is ready to work on his dissertation research. Alvarez is disappointed.

It appears that there was no discussion, persuasion, or consent between Jonas and Alvarez, that is, none of the kinds of interactions that build trust and forestall misunderstandings. Alvarez's disappointment might lead to a dampening of his enthusiasm and inclination to put forth new ideas. To the extent these results occur, both Alvarez and the research group might lose out. More important is the question of the fairness (on some plausible criterion of fairness) of Jonas's apparently failing to see to it that Alvarez gets proper credit for the idea he contributed. He should at least be assured that for proposing the idea he will receive readily visible acknowledgment in any report or publication that issues from the research. Recognition is the coin of the realm in science. To be denied credit for a contribution important enough to become a research problem for a member of the group is to be denied what is owed.

The broader question to be addressed concerns the ownership of ideas in the research group. The act of taking Alvarez's idea from him and handing it to Wright implies that a communal ownership of ideas prevails. (We assume that Jonas is not an arbitrary dictator.) Should those terms apply to everyone in the lab, senior and junior people alike? If so, would Jonas give away her own ideas? If the terms of sharing are different for junior investigators, what is the basis? Can that rationale be justified? That is, is it compatible with agreed upon ethical standards and are the consequences of the arrangement acceptable?

In the vignette that follows, the broader competitive environment in which scientists operate makes itself felt in the relationship between a supervisor and a postdoc. In addition to presenting a problem in a relationship marked by a disparity of power, the situation poses the more general question of what ethical constraints there should be on competition in science.

Vignette 3: Submitting An Abstract[23]

The deadline is fast approaching for submitting abstracts for a specialized conference in the neurosciences. The postdoc, Jay Patel, has collected data for a project that theoretically promises very exciting results. So far, however, the data from an initial study are not very interesting. In view of the importance of the conference and the promise of the project, Patel's advisor, Helmut Braun, judges that they must submit an abstract. With that aim in mind, Braun looks over the data, fills in missing data points, and eliminates others, explaining that they are noisy. A statistical recalculation using the altered data supports Patel's hypothesis.

However, the unaltered data look satisfactory to Patel. He, therefore, tries to persuade Braun that the data are sound and should not be altered (even if that means they have to miss this deadline), but Braun refuses to budge. Braun insists that Patel send in the abstract for their co-authored paper and include the altered data. He points out that they can always withdraw from the conference if more definitive experiments do not pan out. Braun says he knows of other occasions when papers have been withdrawn from conferences after abstracts were accepted. Of course, the result was a hole in the program.

Is there anything to be said in defense of Braun's conduct? The best that can be said is that he believes the data points he has added will eventually be derived from Patel's further investigations and that the data points he has eliminated as noise will prove to have been just that. It is to Patel's credit that he does not lose his ethical bearings; he is troubled by the dishonesty of submitting an abstract with the altered data. Patel, however, is under intense pressure from Braun, who is apparently anxious about his position in a highly competitive research environment.

Should Patel be persuaded by Braun's observation that they can withdraw from the conference if further experiments fail to yield the expected results? After all, Braun seems to suggest, they do no harm so long as they do not present or try to publish results based on altered data. (Of course, if the abstract is to be published in a program book, this rationalization is not available.) Patel may wonder whether this anticipation of research results is common practice in science. He might even ask himself whether this could be a convention in this field of science. Though he is moved to entertain such questions, Patel may be embarrassed to inquire about the propriety of Braun's conduct.

The reasons against having such a convention or accepting Braun's rationale for going along with submitting the altered data coincide with reasons for the moral prohibition of deception generally. Social enterprises depend on trust. To the extent that we cannot rely on people to deal truthfully, we have to be wary and may even have to take defensive measures. When people find it necessary to take such measures, they have to engage in costly efforts that impede their enterprises. It would be very difficult to plan and conduct cutting-edge research conferences if those who submit abstracts cannot be counted on to be truthful. Braun is counting on general honesty and making an exception of himself and Patel; he is prepared to take advantage of the honesty of others.

Braun has created a very difficult situation for Patel. He is setting an example of dishonestly cutting corners; he is leaning on Patel to behave dishonestly; and perhaps he is leading him to wonder what other unethical behavior Braun will require as Patel continues to work with him. Braun is also exposing Patel to added risk. Braun cannot be certain that the only untoward consequence they might have to face is the necessity of withdrawing the paper. This is a situation in which the ethical problem for Patel becomes one of figuring out how to avoid involvement in submitting an abstract containing "fudged" data, without alienating his advisor.

Patel needs to find someone to talk to. Someone within the research group who has won Patel's trust (if there is such a person) would be the first candidate. Patel may have to go outside the group; pressure to misrepresent warrants that action. In a large research university, there should be a multiplicity of channels for questions and complaints and there should be, as an ultimate resource, an office designated to deal with such problems. If Patel lacks these options, he will have to turn to the department head or another experienced person with authority who has inspired trust. In telling his story, Patel should make every effort to stick to a narrow factual account. That is essential, both ethically and practically.

V. GENERAL DISCUSSION OF ISSUES FROM VIGNETTES

The three vignettes comprise a sample of problematic situations that researchers report. Though they merely open a window on difficulties that arise, they highlight some issues that must be addressed in managing research groups, and suggest standards that are needed to maintain a research environment that supports ethical conduct.

First is the necessity for articulating ground rules. Second is the importance of clearly assigning and demarcating research problems for graduate students and postdocs. Third is the need to devise regular channels for communicating and sharing information concerning techniques, data, analysis, and interpretation. Fourth is the necessity for formulating clear policy about the basis and opportunities for recognition. Fifth is the need for stating clearly a policy about criteria for credit and authorship, with the rationale for each. Sixth is the importance of devising clear policies concerning control and ownership of data or ideas.

We have already noted, in connection with Vignette 1, that research groups in universities have latitude for devising alternative sets of ground rules, for example, rules about retention and ownership of data. For research groups in companies, constraints specific to the world of commerce may narrow the range of alternatives. Even in companies, however, deliberation within research groups about alternative ground rules can be beneficial, producing at least a clearer sense of what is at stake from an ethical perspective.

The vignette of a graduate student's surrender of a thesis idea, Vignette 2, brings up the concept of the "toxic mentor." Scholarly discussion about toxic mentors has distinguished four toxic types: (1) "avoiders," (2) "dumpers," (3) "blockers," and (4) "destroyers or criticizers."[24] By making themselves unavailable to students, by dropping students into new roles to "sink or swim," by obstructing with delays or other means, and by tearing down with criticism, advisors can cause unnecessary pain and impede students' progress.

Unfortunately, this classification of four toxic types is not exhaustive. We can add other types: advisors interested only in making clones of themselves, those who overprotect, connivers or manipulators, and those who continually change the playing field so that students do not know what the rules are. Another category includes advisors, like the one in Vignette 3, who transmit messages that cutting corners, misrepresenting, and free riding on the ethical behavior of others is acceptable. The situation in Vignette 2 (Whose Thesis Problem?) shows one of the many ways in which an advisor can do damage in launching a student with a problem suitable for a dissertation. The damage can be severe enough to drive very talented students out of graduate study.

Vignette 3 brings out problematic aspects of the position of the postdoc. One of the positive features of the postdoc position is that it forms a bridge between graduate students and full-fledged investigators and thereby has the potential to enhance the research milieu. In that bridging position, the postdoc can stimulate or inspire graduate students and contribute to the thinking and work of more senior people in the research group. That bridging feature can, of course, benefit the postdoc, tying the trainee in the most tenuous position more firmly to the group.

In view of the complexities of the postdoc's situation, it takes considerable thought and care on the part of the postdoc's supervisor to work out an arrangement for fully realizing the potential of the postdoc, not to mention to ensure that the postdoc is neither neglected nor exploited. It must be emphasized that realizing the benefits requires planning and care. An example of a mechanism that has worked is the creation of a small seminar, led by the postdoc, that concentrates on the postdoc's specialty and includes graduate students and more senior investigators who have an interest in that specialty. When the advisor takes the trouble and arrangements work out, the benefits for everyone can be striking.

At the Sigma Xi Forum on Ethics, Values, and the Promise of Science in February 1993 in San Francisco, a panel of postdocs poignantly conveyed the conflicts, pains, and precariousness of the postdoc position.[25] Trainees in that position can labor under excessive demands costly to them, but of benefit to their advisors or graduate students. Because of the precariousness of their positions, postdocs have almost no recourse when arrangements for them do not work out. They are entirely dependent on the support of their individual advisors, except in unusual cases. One panelist at the Sigma Xi Forum proposed the following guidelines for advisors in their relations with postdocs.[26]

1. Discrimination, Sexual Harassment, etc.: These behaviors are not permissible in any job situation.
2. The Project: A prospective postdoc should have an accurate picture of what the project will entail, within reason . . .
3. Support for the Project: Sufficient resources should be provided for the project. . . . A postdoc should not be asked to wait eight months for the arrival of a piece of equipment which is absolutely essential to a project . . .
4. Group Duties: The level of group responsibilities varies greatly from group to group. The extent of these duties should have been explained clearly and their execution should leave sufficient time so the postdoc can do his/her own research.
5. Project Success: If a project is failing miserably, do not simply assume that the postdoc is at fault. . . . Do not force a postdoc to spend two years working on a bad project in order to avoid admitting that the original idea was at fault . . .
6. Hiring/Termination Flexibility: Both incoming and exiting postdocs need flexibility. . . . Also if your financial situation precludes flexibility, inform the postdoc as soon as possible . . .
7. Work Hours: While it is reasonable to expect that a career scientist will be willing to work more than the arbitrary 40 hours per week, it is not reasonable to expect them in every evening and weekend. . . . Also, flexibility in work hours is a valuable fringe benefit which can be offered at no cost to you.
8. Be a Mentor: Give encouragement and praise where it is appropriate. Help them get jobs. Teach them to be good researchers. Your students and postdocs are your "descendants" and through them your contributions to science and your reputation will continue long past the end of your career.
9. Project Direction: As the postdoc becomes familiar with the project, give them more control. They are Ph.D. scientists . . . and you will probably achieve better results if you collaborate with them rather than using them as technicians.
10. Acknowledgment of Work: Have a clear policy about how contributions are acknowledged and discuss it with your postdocs. If a postdoc contributes a truly original idea or suggests an original project, consider al-

lowing them to publish independently from you. Do not simply adopt their ideas and write them into future grant proposals and papers.

The panelist at the Sigma Xi Forum believes these guidelines are generally "understood" in academia but "are not universally practiced and are rarely if ever discussed." She suggests that individual departments issue such guidelines and offer "mediation/counseling" when postdocs and advisors develop problems.[27]

An important issue that we have not illustrated as such is the problem of the treatment of women and members of other groups heretofore underrepresented in science. In view of the very low representation of these groups at every level, but especially in positions of power, this is a very pressing issue. Since it is closely associated with issues arising from the disparity of power, the treatment of women, African Americans, and other groups previously excluded can be accommodated in discussion of vignettes. For example, if a senior scientist adds his name as author to his female postdoc's paper after the postdoc completes her work, the junior woman may wonder whether her advisor would have the nerve to add his name to the paper of a male postdoc. There are many men who would say from experience that he would. The situation of women, however, is such that they very naturally are led to entertain that question and cannot assure themselves that power would be exercised in the same way over men.

There are, of course, grosser manifestations of discriminating treatment, some ranking as sexual harassment. The need to take measures to avoid particular advisors with reputations for making the going difficult for women and minority students adds to the strain of graduate study. Even an environment where the main problem is invisibility or exaggerated visibility can be very stressful and harmful.

Our recommendations for change in the management of research groups emphasize openness, explicitness, and back-and-forth communication; hence, they favor more democratic policies. At the same time, the examination of problematic situations underscores the need for leadership by principal investigators and research group directors. They should take responsibility for exercising authority and for setting forth and implementing policies that govern the conduct of research. These are not the only institutions facing the imperative to devise less autocratic practices while maintaining a structure of authority. Researchers and, especially, research directors should accept an ongoing responsibility to scrutinize and modify practices in the light of this imperative.

Notes

1. D. S. Greenberg, "Advisor in the Gallo case calls for reopening probe," *Science & Government Report* 24(9)(1991):1.

2. Greenberg, ibid, 4.

3. National Academy of Science, *Responsible Science: Executive Summary* (Washington, D.C.: National Academy Press, 1992).

4. National Academy of Science, ibid, 1.

5. S. Lyall, "Human relations prove a better subject than the social life of insects," *New York Times* (Sept. 1, 1993):B2.

6. The remark is from a personal conversation between one of the authors and Don Buzzelli, a federal government official in an office at the National Science Foundation that deals with allegations of misconduct in science.

7. We have extracted or constructed the cases from the reported experience of senior and junior investigators, postdocs, and graduate students from a range of fields. We have analyzed many of the cases in a monthly sack-lunch discussion with a group of 12 to 20 faculty from a range of fields in our own university. These faculty have confirmed that the cases present characteristic problems that arise in research groups.

8. M. Barinaga, "Labstyles of the famous and well funded," *Science* 252(1991):1776–78.

9. David L. Hull stresses this view in his book *Science as Process: An Evolutionary Account of the Social and Conceptual Development of Science* (Chicago: University of Chicago Press, 1988).

10. D. Goodstein, "Scientific elites and scientific illiterates," in *Ethics, Values, and the Promise of Science. Sigma Xi Forum Proceedings*. February 25–26 (Research Triangle Park, N.C.: Sigma Xi, 1993), 61–75.

11. For a vivid description of effects of the power disparity, a postdoc offered the following account:

> . . . several important issues that I think are common to all postdocs. The first is there are very few options available to postdocs when there is a true conflict. It seems like the only two that are really feasible are suicide or luck. That's pretty much it. You can have either social suicide by completely rebelling against your lab or you can have professional suicide. It's not only the postdoc who is right by any stretch of the imagination, but it seems like the suicide always works the same way. It's very rare that a lab is destroyed by a postdoc where there are many postdocs who are destroyed by labs. ("Postdoctoral researchers: A panel," Ethics, Values, and the Promise of Science, 47–59).

12. M. Davis, "Ethics across the curriculum workshop: Handout," Center for the Study of Ethics in the Professions, Illinois Institute of Technology, Chicago, Ill., 1994. This formulation draws upon Bernard Gert's analysis in *Morality: A New Justification of the Moral Rules* (New York: Oxford University Press, 1988). See also Chapter II of this volume.

13. Davis, "Ethics Across the Curriculum."

14. See Caroline Whitbeck's letter to the editor of *Science* (265[5275] [August 19, 1994]:1020) for a description of competition between graduate students within research groups. Sheila Jasanoff comments on the "winner take all" feature of the organization of science in the United States in "Innovation and integrity in biomedical research," *Academy of Medicine* 68 (9 Suppl.) (Sept. 1993):S91–95.

15. Gerald Seltzer, of the National Science Foundation, in e-mail communication to the authors.

16. M. S. Anderson, K. S. Louis, and J. Earle, "Disciplinary and departmental effects on observations of faculty and graduate student misconduct," *Journal of Higher Education* 65(3)(1994).

17. This point is drawn from an analysis by Alan Wertheimer presented in a paper at the annual meeting of the Association for Practical and Professional Ethics, Arlington, Va., March 2, 1995. For probing discussions related to exploitation and its moral implications, see Alan Wertheimer's work on coercion, for instance, "Coercive offers," a paper presented at the Eastern Division of the American Philosophical Association, Boston, December 30, 1994, and *Coercion* (Princeton, N.J.: Princeton Universeity Press, 1987).

18. Judith Swazey makes these and many other useful observations in "Advisors, mentors, and role models in graduate and professional education: Implications for the recruitment, training, and retention of physician investigators," an unpublished background paper for the Institute of Medicine, Washington, D.C., 1993.

19. J. Henderson and M. W. Olga, "Mentoring in higher education and industry: Is there a paradox?" Paper presented at the Annual Meeting of the American Educational Research Association, Atlanta, Ga., April 1993.

20. At a Conference on Mentoring in Chicago on March 31–April 2, 1995, directed by Professor Robert Sprague of the University of Illinois, with funding from the National Science Foundation, Caroline Whitbeck reported on a program developed by MIT to prepare graduate students to ask questions they need answered about their graduate study. At the same Conference, Michael Zigmond of the University of Pittsburgh reported on the workshops and short courses in "survival skills" he has organized at the University of Pittsburgh to ensure that graduate students receive the information and advice they need for building careers as responsible researchers.

21. The case that follows was produced in 1993 by a research group at the Illinois Institute of Technology in the course of considering its own ground rules.

22. This vignette was written by Weil and Arzbaecher and is based on an actual incident discussed by the IIT Research Ethics Sack Lunch group.

23. This case is adapted from a case in a collection of cases circulated by Leslie Rothenberg, University of California, Los Angeles.

24. L. A. Darling, "What to do about toxic mentors," *Nurse Educator* 11(2)(1986): 29–30.

25. A. L. Singer, G. Jones, J. Gurley, L. Backus, and T. Meyer, "Postdoctoral researchers: A panel," in *Ethics, Values, and the Promise of Science,* 47–59.

26. Ibid.

27. Ibid.

CASES FOR CONSIDERATION

CASE 1. ENDING CO-AUTHORSHIP[1]

Helen Mather is a junior faculty member in the chemistry department she entered as a postdoc five years earlier. She continues to work in an area closely allied to that of her mentor, Henry Goldberg. They have collaborated on a number of papers including a well-respected review article. Mather is concerned that her contributions to the field are overshadowed by Goldberg's reputation so she is especially pleased to be invited to present recent work at an international meeting and to contribute a manuscript to the proceedings.

When she receives the galley proofs of the article she has submitted, she is surprised to find that Goldberg's name has been added as second author. From the departmental secretary, Mather learns that Goldberg had seen the paper on the secretary's desk just before mailing and had then added his name. When Mather asks Goldberg about it, he apologizes for forgetting to mention it to Mather earlier. Mather is upset. While the work described in the paper and presented at the meeting uses a technique Mather and Goldberg had developed together and builds on previous work they had done in collaboration, the new data presented, the figures, and the text were done by Mather independently.

QUESTIONS FOR DISCUSSION

1. What is ethically at stake in this situation?
2. Why is authorship not a trivial matter?
3. What would be a good policy for such situations?

CASE 2. A PROBLEM IN STATISTICAL CONSULTING ETHICS[2]

Background

A Wright State University graduate student consulted with Dr. Kathleen Beal, one of the consultants at the Statistical Consulting Center, about how to analyze her data. The design was straightforward: a three-factor repeated-measures ANOVA (analysis of variance), with repeated measures of one of the factors, and four subjects randomly assigned to each of the 20 factor-level combinations involving the remaining two factors. The graduate student carried out the analyses herself and subsequently defended her thesis successfully.

After the student graduated, her advisor submitted a manuscript on the thesis work to a peer-reviewed conference proceedings, including the student's name as one of the co-authors, but not giving her the opportunity to review the manuscript before submission. The manuscript was conditionally accepted by the editor; the statistical part of the manuscript was not questioned at all. So, it was a simple matter of revising the manuscript according to the referee's comments and resubmitting for final approval.

Problem

Before resubmission, the student had the opportunity to review the manuscript and noticed that her advisor had rerun the analyses without including the SUBJECT term in the model, which led to a declaration of significance for more of the model terms, and hence more interesting conclusions, than appeared in the thesis. The student thereupon insisted that her advisor replace the analysis in the manuscript with the analysis results from the thesis (which included the SUBJECT term in the model, but which resulted in fewer significant effects). The advisor refused to do this, threatening to remove the student's name from the co-author list on the manuscript. The student consulted with the Chair of her department and a senior faculty member and finally came to us asking that we intervene. How should we respond to the request?

Questions for Discussion

1. What ethical issue(s) are involved in this conflict?
2. How could explicit policies and channels of communication reduce the likelihood of such conflicts?
3. Consider and evaluate some options for dealing with this conflict.
4. What advice should the consultant offer?

Notes

1. This vignette is adapted from a case in a collection compiled by C. K. Gunsalus, Vice Chancellor for Research, University of Illinois at Champaign-Urbana.

2. The source of this vignette is *The Statistical Consultant,* American Statistical Association, 11(2)(Summer 1994):2, 3.

Fight Over Data Disrupts Michigan State Project

ELIOT MARSHALL

The university's handling of a dispute between a graduate student and her former professor has upset faculty members.

There are many ways for the partnership between a Ph.D. candidate and professor to come undone, but a recent case at Michigan State University—in which a frustrated graduate student allegedly seized and removed materials from her professor's lab—reveals how intense the struggle for academic credentials can become. It also shows, according to the professor, how timid university officials can be when confronted with an angry student. In this case, the professor claims he's been trying to recover the "hostage" lab data for 18 months, with minimal help from MSU authorities. The student is still in the graduate program at MSU and is trying to publish an article based on the disputed data.

This account is necessarily one-sided because neither the student nor university officials would discuss the situation with *Science.* The professor also skirted some details, noting that in November he filed a formal charge against other members of the faculty, which is the subject of a preliminary inquiry due to end by January.

However, Associate Dean Justin McCormick, now serving as one of the student's academic advisers, points out that the university has an obligation to protect student interests as well as faculty prerogatives. He describes the situation as a divorce in which the parties are seeking a fair distribution of joint property. And he comments that the professor has become "obsessed" with the case.

But this is the kind of dispute that triggers obsessive behavior. In addition to raising questions about academic responsibility, it presents some broad ethical issues as well, such as: What happens when partners in a collaborative project split up? Can one of them use jointly produced data if the others object? Can an adviser dismiss a student and keep the data he or she has produced?

These questions arise out of a conflict that has been raging at MSU's Colleges of Veterinary and Osteopathic Medicine since May 1989, when a senior professor of microbiology named Jeffrey Williams, described by former students as an excellent but "demanding" teacher, dismissed his Ph.D. candidate, Maie ElKassaby. ElKassaby then removed the tissue samples and data on which she had been working, according to present and former faculty members.

Williams considers these materials the common property of the "Sudan Project," an international parasitology project funded by the National Institutes of Health, directed by Williams in collaboration with physicians in the Sudan and researchers at the Upjohn Company. Indeed, Williams did persuade ElKassaby to return some of the material to him, but not all. And in a recent strange twist, the university obtained the rest of the missing data, but has refused so far to turn them over to Williams.

Five university officials interviewed by *Science* declined to comment on the situation, and ElKassaby, contacted by telephone, had nothing to add, declaring it a "private matter" and directing questions to Mary Elizabeth Kurz, the university's legal counsel. Kurz said she cannot discuss specifics because the law requires her to protect the privacy of students. Invoking a legal catch-22, she also declined to discuss the generic problem of data ownership because, she said, one must first know the specifics of a case. But members of the microbiology and pathology departments, and others who have left MSU, provided documents and background information. Many, in fact, are concerned about the university's posture, perhaps fearing that they, too, could be put at the mercy of a litigious collaborator and receive no help from their university.

Although the quarrel hasn't made news locally, it has upset the affected departments at MSU. James Jensen, a former MSU professor in the Sudan Project now at Brigham Young University, says one of the reasons he left MSU was that he became "disgusted" by the way officials are handling this problem. Robert Garrison, a former student in the MSU project, now at Purdue, says the same. "The university screwed this up from the beginning," he adds. He thinks university officials "just turned their back on Dr. Williams."

Meanwhile, members of the microbiology department have been asking the university to clarify its position. MSU Provost David Scott has received two petitions asking for an explanation of what's going on—one in October signed by 24 faculty members and another in November signed by 28 graduate students. Scott has promised to meet with the faculty soon, but at this writing he hasn't responded to the students' letter or set a date for meeting with the faculty.

Among the petitioners are sources who say that Williams and ElKassaby had loud and angry disagreements over how to conduct a part of the research project, and they report that ElKassaby had failed a preliminary exam in her departmental field, pathology, before Williams dropped her as a student. According to some graduate students, ElKassaby took two actions following her dismissal: She filed a grievance against Williams, reportedly charging that Williams had acted arbitrarily and without warning, and she removed data she felt belonged to her.

According to a five-page statement filed by Garrison after he left for Purdue, ElKassaby also told MSU officials that she had been "cheated out of a patentable invention by Dr. Williams and scientists at the Upjohn Company." Garrison claims that the university has already established this claim of ElKassaby's to be "groundless." Two top research officials at Michigan State, former vice president John Cantlon and associate vice president Henry Bredeck, looked into the charge in 1989 and eventually concluded it had no merit. After weeks of probing in which the accused were not notified directly of the inquiry, Cantlon and Bredeck wrote letters of apology to both Williams and Upjohn. Cantlon's letter to Upjohn, dated 21 December 1989, said, "This inquiry identified no evidence to support any mishandling of intellectual property by Dr. Williams or any of the researchers or staff in the Upjohn Company. On behalf of Michigan State University let me express our regret over any discomfort or concern that an inquiry caused Upjohn personnel." Cantlon has since retired and could not be reached, and Bredeck had no comment.

After a period of quiescence, the dispute escalated this summer when Williams learned that ElKassaby, still in MSU's graduate program, intended to publish an article based upon her work in the Sudan Project. According to Garrison and Jensen, ElKassaby thought that the data she produced were hers to use. But Williams insists that she must obtain permission from other collaborators in the Sudan Project. Not surprisingly, he and the Sudanese have reportedly refused to grant it, and now question their very validity.

Nevertheless, the university appears to be willing to allow the publication to go forward, according to Williams and other observers. This posture may have caused Williams to decide to go over the university's head. Recently, he called on the police to intervene and recover the material, which he felt he could not get in any other way. But university officials informed the police there was no crime to investigate because the missing material had been surrendered to Associate Dean McCormick. He told the police they would be returned to the Sudan Project after ElKassaby has signed a legal release. This infuriated Williams, and is undoubtedly the basis of his formal charge of scientific misconduct against the university officials.

Failing to receive satisfaction from the university, Williams has also sought intervention by NIH's Office of Scientific Integrity. This is not the first time he attempted to bring NIH into the case. Last summer, Williams tried to get the NIH institute that funded his work to investigate what he views as misappropriation of data. But in a letter dated 6 July, 1990, John R. La Montagne of the National Institute of Allergy and Infectious Diseases informed Williams that the case "falls within the jurisdiction of local law enforcement authorities and the university itself."

This logic seems to support the university actions. Vice president for research Percy Pierre told *Science* that any data produced under contract with the faculty belong to the school. Since the university now has possession of the data, officials may feel the fight over custody is moot. Pierre wouldn't say, but he forecast a resolution soon, promising to disclose more at that time.

One big loser in all this may be the Sudan Project itself. When one industry scientist says, "This is not trivial research," he has a point. The work that involved ElKassaby focused on a drug called ivermectin, recently adapted for use in humans to treat onchocerciasis, a parasitic disease in the developing world known as "river blindness" that is estimated to infect about 17.5 million people. ElKassaby had been asked to test a radioiummune assay that would detect low concentrations of ivermectin in blood and tissue. The Upjohn Company donated free of charge a testing protocol and radiolabeled test chemicals. They were hoping to use the information to develop a general pharmacokinetic model for antiparasite drugs. Physicians in Mexico and the Sudan played a major role, contributing human tissue and blood samples from people infected with the worms that cause river blindness. Jensen added that it's not easy to get such material from the Sudan: "There were three coups, two civil wars, and three famines."

But now an uncivil war on a U.S. campus has shut down a laboratory that had survived all that. Williams—who won an MSU distinguished professor award in 1982 and holds the 1979 Henry Baldwin Ward medal for parasitology research—has decided to end the project at MSU after

11 years and take early retirement. He says he is "disillusioned" by the way officials failed to support his claim to data from his own lab. The grant is being transferred to Brigham Young University.

"The real losers" in this dispute, Jensen says, are the Sudanese, who made a great effort to collect the samples and whose claims "are being ignored." At the same time, some of the faculty at MSU fear that unless universities learn to handle such conflicts better, collaborative research on campus will become a risky proposition.

5

Conducting, Reporting, and Funding Research

A: CONDUCTING RESEARCH[1]

STEPHANIE J. BIRD AND DAVID E. HOUSMAN
with a scenario by Eve K. Nichols

I. INTRODUCTION

LATE ONE NIGHT[2]

(After a group meeting on Tuesday afternoon, Professor Barbara Steel speaks with postdoctoral fellow, Sandra Dunn.)

PROFESSOR STEEL: Sandra, you were unusually quiet at group meeting today. I thought you'd planned to discuss the results of your last fractionation. I wanted to go over the data with you this morning, but when I checked at your bench at eleven o'clock you hadn't come in. Is something wrong?

SANDRA: No, nothing's wrong. I was reading the gels late last night and I overslept. I have a meeting now outside the building, but I'll knock on your door when I come in tomorrow.

PROFESSOR STEEL: I'll be here but try to catch me before lunch. I have appointments most of the afternoon.

(Three days later, Professor Steel stops her graduate student, John Palant, in the hallway.)

PROFESSOR STEEL: John, have you seen Sandra? She said she'd stop by on Wednesday to go over her data with me, but I haven't seen her since group meeting.

JOHN: She hasn't been around much during the day, but I know she's been working at night. You know, it's strange. Monday she said she had an idea that might help me find the co-activator for my DNA binding protein. I asked her about it at the meeting, but she said she'd been wrong and I should forget about it. I've been so frustrated the last few weeks that I haven't been coming back in after dinner.

PROFESSOR STEEL: I know it's been hard, but I'm sure you're on the right track. You found the DNA binding protein; you just need to find the co-activator to make the whole thing work. The changes we discussed at group meeting might do the trick. I've got a committee meeting now. Will you leave a note on Sandra's desk asking her to call me?

JOHN: Sure. I'll let you know on Monday how things worked out.

(Monday morning in Professor Steel's office there is a knock at the door.)

PROFESSOR STEEL: Come in. Oh, Sandra, it's you. I've been trying to reach you for three days. Where've you been?

SANDRA: Take a look at these. (*She hands Professor Steel some papers.*)

PROFESSOR STEEL: What are they?

SANDRA: I've drafted two papers. One describes the work we planned to talk about last week. I realized when I read the gels last Monday that I'd accidentally found the answer to John's problem. Suddenly it was clear that we had an entirely new class of DNA binding proteins and their partner-co-activators. I just needed one more experiment to confirm the results.

(Professor Steel quickly reads through the two papers.)

PROFESSOR STEEL: This is terrific; I can't believe we didn't see this before. But Sandra, what about John? Why didn't you tell him you'd found the answer to his problem. I mean, this is his thesis project. You could have done the last experiment together. He should be included in the final paper, too.

SANDRA: I don't think so. I've thought about it a lot. I put his name on the first paper because I started with his technique for isolating the DNA binding activity; but the second paper on the co-activator and its implication for all regulation is mine. I want it to stand out in the journal with just two authors.

PROFESSOR STEEL: I can't force you to put John's name on the paper but I think you should consider it again. I like to think we all work together in this lab. Have you shown these papers to him yet?

SANDRA: No, I thought I'd present them at group meeting tomorrow. What do you think?

This scenario raises a number of ethical issues including intellectual property and the ownership of ideas, authorship, professional relationships, and the responsibilities of mentors. It also suggests that there are underlying pressures and varying perceptions of professional values and standards that influence behavior.

This two-part chapter is concerned with the ethical issues that arise in the course of carrying out and reporting scientific research. Although we will identify some of the ethical problems associated with common practices and conventions in conducting and reporting research, the solution to the problems may not be readily attainable. The competing needs and claims of the parties involved and the broader implications of alternative courses of action may be unclear.

In this chapter we examine the various stages of research from hypothesis development and research design, through data collection and handling, to the reporting of research findings within and beyond the scientific community, and the funding of research. Each step in the development of a scientific study requires decisions. Choices must be made that will bear on the ultimate outcome of the study, as well as the way in which it will be perceived by the scientific community and society at large. These choices must be made within the framework of value systems held by the scientists carrying out the study with respect both to scientific conduct and to their roles as members of society at large. Here we examine the ethical concerns that arise throughout the process of conducting and reporting research.

In addition to various personal and professional values of the individual and of society, a number of external and internal pressures affect research practices. Some pressures come from the availability of funding. Others are applied by the scientific community in the form of the state of current thinking within a particular discipline, the expectations of colleagues, and the real or apparent competition among colleagues. Still other pressures are generated by the individual as a result of personal or professional expectations, goals and ambitions.

II. INTELLECTUAL PROPERTY

An element affecting the ethics of conducting and reporting research is the professional needs of the scientist. The major source of professional capital for a scientist is the product of his or her intellectual energies in developing hypotheses, designing critical tests of those hypotheses, interpreting the data obtained in those tests, and developing new lines of investigation. The scientist's need is, therefore, to be recognized for the contributions which she or he has made to science. This is embodied in the concept of ownership or rights to intellectual property. However, rules and conventions as to who legitimately should receive credit for scientific findings interact with both the value system governing the conduct of scientific research, and the values that concern the impact of scientific work on society at large.

Frequently, inherent or potential conflicts can arise between the necessity to establish intellectual property rights and other values. For example, the need for scientists to receive critical evaluations of their scientific thoughts and findings from peers runs the risk that scientific credit for the work may ultimately be shared with, or lost to, those with whom critical information is shared, either intentionally or unintentionally, at an early stage of the work.

Within a research group, the way in which ascribed scientific credit will be understood by outside observers often is a concern that creates significant conflict, particularly if the rules regarding assignment of such credit are vague.

The needs of funders to receive information critical to the goals of the funder, and to the overall good of society in a timely manner, present another aspect of the issues that arise from the notion of intellectual property. From the funders' perspective, research results are a consequence of the funding process that made the work possible and, as such, are sometimes considered the property of the funders rather than the scientists who carried out the work. However, research findings are also the result of the intellectual efforts of individuals who expect to receive appropriate recognition and compensation for their creative contributions. The potential for application of research findings to pressing social problems, with possible concomitant commercialization, exacerbates the tension between funders and researchers, and their competing claims for control over intellectual property as it relates to scientific research are a critical, complex, and evolving issue.[3]

III. RESEARCH DESIGN

A hypothesis reflects an attempt to logically determine the cause and effect relationship of observed phenomena. It is then usually tested for its predictive value. In disciplines where predictions cannot be readily tested (e.g., geology and astronomy), good hypotheses unify widely differing findings. Good hypotheses need to be internally consistent; simple hypotheses are viewed more favorably than complex ones.

The generation of a hypothesis is an essential first step in any scientific endeavor. What ultimately drives a scientist's choice of hypothesis is a combination of interest, a knowledge of previous related work, analysis of current data, and internal conjecture about the problem. A scientist who sees the problem in a unique way can significantly advance the field with a hypothesis that breaks new ground, such as the revolutionary hypothesis that the earth revolves around the sun. A new hypothesis may also be the inevitable necessity of the outcome of a previous study, which leads to a set of findings that do not have a clear explanation within the current framework. A hypothesis that attempts to explain the new findings now becomes both a scientific possibility and necessity.[4]

The selection of a hypothesis can be especially vulnerable to the unconscious injection of values not generally considered scientific.[5] Because the way one frames a research question is related to one's perception of the world, researchers may bring philosophical, cultural, religious, political, or economic values to their formulation of a hypothesis. This is not surprising since science is carried out in a culture that is incorporated into the experiences and world view of scientists long before they begin to think in a scientific framework. However, to the extent that social values are unconsciously or surreptitiously incorporated into the development of a hypothesis, ethical issues arise since a dynamic relationship exists between hypothesis, observation, and interpretation. Values that underlie an inequitable policy or system may receive added credibility because of the misperception that the scientific process is free of social values. Indeed, policies may receive additional weight because of the imprimatur of objectivity, highly prized in Western culture.

The choice of hypotheses has practical determinants as well. The hypotheses one chooses to test are governed by the techniques one knows, the facilities to which one has access, and the financial, material, and personnel resources that are available. As funding usually plays a crucial role in determining which hypotheses will be tested and which will remain conjectures, the process of garnering economic support is one that inevitably involves ethical elements. In order to obtain financial support, a scientist must in some way justify the relevance of the scientific work to be carried out, since society itself must justify expenditures for scientific research in a world of many and important competing claims for finite resources. So researchers may be tempted to formulate the research question that they are interested in addressing in a way that exaggerates its relevance to the funded area. By setting priorities and restricting support to research that addresses those priorities, funders disregard the fact that it is often not possible to determine which research area will ultimately have the greatest relevance to a specific area of social concern. Three decades ago, research in bacterial enzymes was not expected to open the door to genetic engineering. Yet molecular genetics and biotechnology have become central to many areas of biomedical, agricultural, and industrial research and a potential key to health, global food production, and industrial development, which were major societal concerns 30 years ago.

The value system of society can also have a strong bearing on the research area that a scientist chooses because of the appeal of a recognized and lauded cause. Why not choose to focus on a cure for cancer as a scientific objective if the President and the U.S. Congress have declared a "War on Cancer," rather than some other area of biological research?

Using procedures and techniques with demonstrated reliability, strategies are developed and carried out in order to determine, and then refine, the predictive value of the hypothesis. Of particular concern is the possibility that the experimenter's conscious or unconscious desires can influence the outcome of the experiment. One of the key strategies for addressing this concern is the scientific requirement that results be replicable by multiple investigators. In the clinical and behavioral sciences, various additional techniques have been developed to limit the influence of unconscious bias in data collection and analysis, including a "double-blind" study design, randomization of experimental subjects, and the use of appropriate controls.

Conscious and unconscious assumptions are the basis for generalizations from the experimental case or model to other similar instances of interest. However, the significance of differences between the conditions of the controlled experiment or the features of the model and the "real-world" case need to be determined. In the life sciences, we often are tempted to assume that studies in an animal model are applicable to other species, that research done in one sex (usually males) is applicable to the other, that findings in the adult can be extrapolated to the juvenile or aging animal. These assumptions need to be acknowledged and reevaluated periodically. For example, a compound like thalidomide safely metabolized in adult female rats, with no effect on developing rat embryos, may be found to be safely metabolized in adult humans and yet prove to cause serious defects in developing human embryos.

IV. RESEARCH PLAN EXECUTION

Data collection, selection, analysis, and interpretation are at the heart of scientific research. In many respects learning to identify the signal in the midst of noise, to distinguish what is significant from what is artifact, and to collate, assemble, and synthesize bits of information into a coherent description of a phenomenon is the purpose of research training.

A. Data Collection

The accurate recording of the observations in a scientific study is a crucial component of its successful conclusion. In the abstract, the recording of all significant observations would seem to be straightforward. In reality, the forms used in recording observations are themselves cultural artifacts. One of the earliest critical experiences of a student is the review of primary data with a supervisor. Inevitably, the notion of what is essential to be included in the collection of data is derived during this socialization process. What is essential or desirable to include in the written representation of observed data can vary considerably from discipline to discipline. In all fields, the implicit presumption is that the data represented in the notes are those that represent the relevant variables in a study. However, if the study is being carried out at the very edge of scientific understanding, the variables that are relevant are not fully appreciated at the time the study is conceived. Part of the insight of a scientist may involve noting variables that others may not perceive as significant and establishing a relationship between these variables and experimental outcome. Thus, the process of scientific observation involves making judgments within a cultural context that asserts its own value judgments.

Researchers clearly have the responsibility to collect data with as much precision as possible. Frequently, the differences between two scientific studies that reach seemingly opposite conclusions can be traced to differences in attention paid to accuracy in collection of the data.

Reliable data collection depends upon consistent techniques. and this takes practice. The more and various the steps, the more time is required, though different individuals bring different skills and experience to experimental technique. During the learning period, it is expected that errors and mistakes will be made. Resources of all types, such as time, materials, and samples, will be invested in the learning process without production of reliable data. Pressures on researchers arise if this investment in training is not included in the research budget and plan, if resources are limited, or if statistical significance is an issue; a conflict of interest may arise in deciding how to use data gathered during the learning process. The pressure to produce results within a specific time frame can lead to the acceptance of marginal data in support of a hypothesis. Time-based deadlines may arise from deadlines for renewal of funding or from a perception that there is a "race" among research groups to reach a particular objective.

Consistent technique requires effort and attention to detail. Sloppiness in handling chemicals, in recording observations, or in calibration of equipment results in inevitable error and mistakes for the experimenter and co-workers. Errors refer to deviations from correctness, things that can be right or wrong, like spelling or copying errors. They are inherent at the limits of measurement and are reflected in standard statements of the margin of error, as in "12.5 inches ± 5%." Mistakes are not only obvious errors but those that result from carelessness, inattention, or misunderstanding and may require time to evaluate.

Avoidable mistakes include overlooked variables such as differences in contaminants on glassware from one time or location to another, or differences in pipetting technique from one experimenter to the next. As a result of a sloppy technique, stock chemicals, a work area, or even a cell line or population of animals can become contaminated. Overextrapolation from a functional model or from research results is also a source of avoidable mistake, for example, in applying findings obtained in research in adults to children, or in women to men.

Avoidable errors include miscalculations, misread scales, recording the wrong number, and applying the incorrect computer program. Needless error can be minimized by checking calculations for accuracy, and by testing calculations with a wide range of variables. Reasoning and underlying assumptions should be checked and rechecked by the researcher and by others, especially those with differing experience and perspective. But, some errors are unavoidable. Not all variables can be foreseen; for example, individual differences in sensitivity to certain chemicals may depend on hereditary, environmental, or cultural factors yet to be identified. In addition, critical experiments may not be socially acceptable because of the risk to human subjects, or the suffering caused to animal subjects. Assumptions that reflect individual or cultural bias can be a source of both avoidable and unavoidable error.

B. Data Selection, Analysis, and Interpretation

The process of moving from a set of scientific data to a conclusion is a significant one for the scientist. In sorting through a data set, a scientist must consider the entire set of measurements or observations. Often, outlying data points are removed from the data set if measurements have been repeated multiple times. Statistical tests that permit the removal of such data points are well established in many fields. However, when the data set is small, when criteria are being formulated for differentiating responses to an experimental perturbation, when there are numerous sources of error, or when experimental techniques are at their outer limit of reliability, every data point is critical. As the number of observations increases, it becomes clearer which stray points are likely to be the result of an irrelevant glitch and which are likely to be a significant subset of the experimental group.

Before dropping any piece of data, a scientist must reflect on the circumstances of each observation to consider its potential significance. Perhaps the outlying data points are the clue to an as yet undescribed phenomenon. It is possible that understanding the basis of the outlier may be more significant to the advancement of knowledge in the field than addressing the original question. It is only by making an informed judgment on these issues that the scientist can go on to report a set of conclusions to the scientific community.

Outliers in research that has public policy ramifications can raise additional concerns. For example, that smoking is associated with the occurrence of lung cancer is a basic conclusion that is robust to statistical analysis. But clearly, many individuals who smoke heavily do not get lung cancer. What is the significance of these individuals? Do they represent a subgroup with particular characteristics or are they predicted by the most appropriate statistical models? How the scientist may choose to address this issue raises an interesting question: Does investigation of the reasons why some individuals do not get lung cancer from smoking undermine the message derived from scientific research that indicates that smoking causes lung cancer—a message that is the foundation for what many would agree are socially beneficial programs?

Data interpretation, like hypothesis development, takes place within a framework of theory and previously established hypothesis. This scientific framework can bias interpretation at the limits of instrumentation, techniques, or theory. For example, the biologist van Leeuwenhoek probably identified the green, chlorophyll-containing, one-celled organism *Euglena* as an "animalcule" because it moved, although it is recognized today as largely plant-like.

In addition, expectation and desire influence the interpretation of observed phenomena. The investment of time, effort, and ego, or the pressures on oneself or collaborators to publish for purposes of career advancement can also motivate researchers to see consistency or support for coherence in the data. A reputation built on a particular theory can make one reluctant to accept or even believe contrary findings. If an investigator believes that research findings may influence public debate and affect public policy and political decisions, there may be conscious or unconscious pressures to perceive data or interpret results in a manner not entirely divorced from their expected social impact, such as the extent to which sexual orientation is biologically determined.[6]

C. Data Retention

The question of who retains ownership of scientific data after it has been produced is one that has increasing significance as scientific data enter the legal arena. There is little question that as a study is being carried out the notes and data that are the essential elements of the scientific study must be at the immediate disposal of the scientists carrying out the study. Funders generally expect that the research institution to which funds were granted will retain ownership of data, while the laboratory where the work was carried out will retain custody. Presumably individual contributors should retain access, but this is a policy that is evolving.

In an industrial research setting it is often common practice to microfilm data and notebooks of all scientists on a weekly basis. Notebooks and data are considered to be property of the company in such settings and are surrendered to the company at the end of employment. But against these policies is the academic practice of considering notes and data to be the personal property of the researcher. Most scientists retain the notebooks and records of their scientific work throughout their lives. To do otherwise would break an intellectual trail that reflects the essence of a scientist's identity. In recent years institutions have been called upon to adjudicate disputes within the scientific community or to respond to outside investigation. Under these circumstances notebooks and data records have acquired the status of legal documentation. In response to such situations, policies attempting to regulate the possession of such documents have been promulgated. It is probably fair to say, however, that such policies are not universally regarded as appropriate or practiced within the scientific community at large. While the expectation remains that, should the occasion arise, an individual will readily make notes and records available, many scientists still regard the personal recording of an individual in training to become an independent scientist as a direct and personal reflection of that person's activities as a scientist, and as such these notes and records should most appropriately remain in the possession of that individual throughout his or her scientific career.

Practically speaking, at many research institutions space is at a premium and storage of slides, gels, graphs, x-rays, counter tapes, and so on may not be possible. At the same time, much data may be stored in computers and on computer disks, rather than in notebooks or other sorts of records. However, viruses and computer failures make the data vulnerable, and it is considerably more difficult to detect and verify the alteration of data on computer disk than in notebooks. Legal motivations for regulating the retention of data may not mesh with the research environment.

Notes

1. Some of this chapter has been previously published: S. J. Bird and D. E. Housman, "Trust and the collection, selection, analysis and interpretation of data: A scientist's view," *Science and Engineering Ethics* 1(1995):371–82.

2. Copyright 1991 by the Whitehead Institute for Biomedical Research. Prepared by Eve Nichols with assistance from Professors Gerald E. Fink, Lawrence E. Susskind, and Robert A. Weinberg.

3. D. Nelkin, *Science as Intellectual Property* (New York: Macmillan, 1984).

4. T. S. Kuhn, *The Structure of Scientific Revolutions,* second edition (Chicago: University of Chicago Press, 1970).

5. H. E. Longino, *Science as Social Knowledge: Values and Objectivity in Scientific Inquiry* (Princeton, N.J.: Princeton University Press, 1990).

6. D. H. Hamer, S. Hu, V. L. Magnuson, N. Hu, and A. M. L. Pattatucci, "A linkage between DNA markers on the X chromosome and male sexual orientation," *Science* 261(1993):321–27; S. LeVay, "A difference in hypothalamic structure between heterosexual and homosexual men," *Science* 253(1991):1034–37; S. LeVay, and D. H. Hamer, "Evidence for a biological influence in male homosexuality," *Scientific American* 270(1994):44–49.

CASES FOR CONSIDERATION

THE NEW TECHNICIAN[1,2]

(Christine Wilson, associate professor in a leading biology department, is on the telephone in her office. She is speaking to Elaine Gardiner, a technician formerly in her lab. As she talks, she idly turns the pages of two different lab notebooks on the desk in front of her.)

CHRISTINE: I'm sorry to bother you, Elaine, but I was wondering if you could drop by sometime this week. I know you have a midterm break soon. (*Long pause.*) Well, when do you leave? . . . Tomorrow morning ... No, you don't have to come over now. You have an exam tonight and you probably have a lot to do to get ready to go. (*Long pause.*) I just thought maybe you could look at Bob's data before the nutrition meetings at the end of the month. His results on the feeding behavior of the elderly rats are—I don't know how to put it—unexpected, I guess. (*Long pause.*) Well, his observations don't show the same correlations between age and behavior that yours did. In fact, some of the variables don't change at all between 6 and 18 months . . . But don't worry, I'll sit down with him and I'm sure we can figure out what's going on. Have a wonderful trip. You've earned a rest going right from the lab here to graduate school with no break.

(Christine hangs up the phone and then picks it up and dials an in-house extension. She is calling Bob Austin, her current technician.)

CHRISTINE: Hi, Bob. Can you come down to my office now? And bring your new lab notebook.

(Christine hangs up and five minutes later, Bob walks in and sits down.)

CHRISTINE: I've been going over your results and comparing them with Elaine's. I don't understand why your results are so different from hers.

BOB: I don't know either. I'm following the procedures she taught me and, before she left, we did several test runs together to make sure I was scoring behavior the same way she did.

CHRISTINE: You know we've been talking about these results in public for almost a year. We've spent three years looking at physiological and hormonal factors involved in feeding behavior and lots of people are waiting for the final results. You're working with the fourth and final set of aging litter mates. If we start another series of experiments on this project now, we'll have to delay the new work on carbohydrate metabolism.

BOB: I'm sorry, Christine. I've checked everything I can think of. The food measurements are straightforward and we're sending the bloods to an outside lab. Maybe you could spend a few days in the lab with me to see if I've missed anything.

CHRISTINE: That wouldn't really do any good, and besides, I have to teach this month. Elaine was always so precise. Maybe you've missed too many observation periods. I told you one or two was OK . . . but you've missed a lot more than that.

BOB: I've stayed within the guidelines that you gave me. Listen, Elaine called me as I was leaving the lab on my way down here. She offered to come over when she gets back from Spain. Maybe that's all we need.

CHRISTINE: That'll be too late for the nutrition meeting. The conference organizers are expecting a draft of my paper in 10 days. Just go back to the lab now. I wouldn't want you to miss another observation.

(Bob turns, clearly upset, and leaves the office. Later that day, Christine joins two friends and colleagues, Rebecca Cohen and Philip Mertz for lunch.)

REBECCA: What's the matter with you? You look like you've had a terrible morning.

CHRISTINE: I really don't want to talk about it. It would just spoil your lunch.

PHILIP: Oh come on. I got the chili pepper special from across the street; I don't think anything could spoil that. Now, what's the problem? I thought everything was going so well.

CHRISTINE: Well, you know that Elaine left to go back to school and I have a new technician. He seemed terrific—his references were great and he picked up the gist of the feeding project very quickly.

REBECCA: So, what's the matter?

CHRISTINE: I've just been going over his data set, and his results are very different from Elaine's. In fact, many of the conclusions that we drew from the first three cohorts don't hold up. I just don't understand it.

PHILIP: Maybe he's not scoring correctly.

CHRISTINE: Maybe, but Elaine checked him out pretty carefully before she left. The frustrating thing is that this fourth cohort casts doubt on all the others. I shouldn't say this, but I wish I'd never started with the fourth cohort. What am I going to tell people at the nutrition meetings?

PHILIP: If I were you, I'd just forget this fourth cohort. Chalk it up to a bad technician and forget about it.

REBECCA: Oh come on, Philip. You really wouldn't do that. You can't ignore a whole data set simply because you don't get the results you expect.

PHILIP: I know, but this is a special case. Christine's first three cohorts were consistent. No one would have expected her to run a fourth set of experiments.

REBECCA: The point is that she did, and the results conflict with her earlier findings. She really needs to find out why. Christine, have you spent time with Bob—that's his name, isn't it—in the lab?

CHRISTINE: That wouldn't really make a difference. Elaine was so good that I really let her take charge of the behavioral study once we'd finished the original design. Of course, we went over it together many times in the beginning, but that was almost three years ago. I've been much more involved with the biochemical studies.

REBECCA: How has Bob handled those?

CHRISTINE: The data are consistent—the behavioral stuff is the real problem and, of course, that's what people are really interested in. You know I get calls almost every day from groups planning to set up human studies based on our data.

PHILIP: Well, I think there's a simple answer, but you're the one who has to make the decision. I've got to get ready for my 1 o'clock class. I'll see you later.

(As Philip leaves, Rebecca looks after him and shakes her head.)

REBECCA: I wouldn't listen to Philip. There's too much at stake here. Your project ties together a lot of work that's been on the books for years. It would be a shame to do anything that might lessen its value at this point.

CHRISTINE: I know that, and I don't really have a problem with my talk at the nutrition meetings. I can explain my concern about the fourth cohort in the oral presentation. The question is, how do I write it up. The Society is doing a special issue of the Journal based on the meeting and I really wanted to have definitive answers. I can't ask them to delay the issue for a year or two while I check out another cohort. The last thing the field needs now is another inconclusive paper. Also, I have two other projects that I've been neglecting; if I don't get them moving, I won't have enough data to go in for the competitive grant renewal at the end of the year.

REBECCA: I know it's difficult, but I don't think you have any choice. I've got to run now, but please call me if I can help in any way.

QUESTIONS FOR DISCUSSION

1. Could Christine have done anything differently to guard against a change in technicians becoming a confounding variable?
2. What is the responsibility of a technician leaving the lab with regard to training a new technician?
3. What should Christine do about the data that doesn't fit with her earlier observations? Under what conditions would it be proper for her to disregard it?

4. How should Christine deal with this data in her talk? In her paper for publication? In conversation with other scientists?

Notes

1. Prepared by Eve K. Nichols and Stephanie J. Bird with David E. Housman.

2. For discussion of this case, see S. J. Bird and D. E. Housman, "Trust and the collection, selection, analysis and interpretation of data: A scientist's view," *Science and Engineering Ethics* 1(1995):371–82.

Science, Statistics, and Deception

JOHN C. BAILAR III

In science, lying is condemned, even by some of its few practitioners. Deliberate or careless deception short of lying, however, seems to be universally accepted and sometimes even promoted as a part of the culture of science. I do not suggest that scientists as a group are careless, venal, or otherwise depraved: They may even be above the human average in developing and adhering to detailed, albeit tacit, standards of professional conduct. Those who are clearly violators are drummed out of our ranks, loudly and publicly. But what about less clear-cut deception?

My thesis is that our professional norms are incomplete and that several kinds of widely accepted practices (Table 1) should also be ideally recognized as potentially deceptive and harmful. Some of these practices also have much value, but at times they are inappropriate and improper and, to the extent that they are deceptive, unethical.

The scientific method is fundamentally concerned with the processes of inference, generally from data that are necessarily inaccurate to some degree, incomplete, drawn from small samples, or not quite appropriate for a specific task. Inference—that is, drawing conclusions or making deductions from imperfect data—provides most of the excitement and intellectual ferment of science. Scientific rewards are probably more closely related to valid inferences established than to such related activities as imaginative hypotheses formulated, elegant experiments designed and conducted, or new methods developed. The rewards for publishing a first-class inference can include income, position and power, professional status, and the respect of colleagues. Such rewards may some times count for more than self-respect and the joy of discovery. We must therefore be

Reprinted with permission from *Annals of Internal Medicine* 104(1986):259–60.

Table 1.
Some Practices That Distort Scientific Inferences

Some Practices That Distort Scientific Inferences
Failure to deal honestly with readers about nonrandom error (bias)
Post hoc hypotheses
Multiple comparisons and data dredging
Inappropriate statistical tests and other statistical procedures
Fragmentation of reports
Low statistical power
Suppressing, trimming, or "adjusting" data; or undisclosed repetition of "unsatisfactory" experiments
Selective reporting of findings

attentive to scientific norms and activities that may distort the processes of inference.

An example of a deceptive practice is the statistical testing (such as the calculating of *p* values) of post hoc hypotheses. It is widely recognized that *t*-tests, chi-square tests, and other statistical tests provide a basis for probability statements only when the hypothesis is fully developed before the data are examined in any way. If even the briefest glance at a study's results moves the investigator to consider a hypothesis not formulated before the study was started, that glance destroys the probability value of the evidence at hand. Certainly, careful and unstructured review of data for unexpected clues is a critical part of science. Such review can be an immensely fruitful source of ideas for new, before-the-fact hypotheses that can be tested in the correct way with other new or existing data, and sometimes findings may be so striking that independent confirmation by a proper statistical test is superfluous. Statistical "tests" are also used sometimes in nonprobability ways as rough measures of the size of an effect, rather than to test hypotheses. (An example is the column of *p* values that sometimes accompanies a table comparing the pretreatment characteristics of patient groups in a randomized clinical trial.) When either the test itself or the reporting of the test is motivated by the data, a probability statement such as "$p < 0.05$" is deceptive and hence damaging to inference.

Other potential problems are the selective reporting of findings and the reporting of a single study in multiple fragments. These practices can obscure critical aspects of an investigation, so that readers will misjudge the evidential value of the data presented. Such reporting may be deceptive, whether deliberately or accidentally. On the other hand, these practices sometimes have positive value that should be preserved. For example, they can facilitate the tasks of both the investigator and the user when a demand for a monolithic analysis might seriously delay or frustrate the progress of both.

"Negative" conclusions of low statistical power—that is, reporting that no effect was found when there was little chance of detecting the effect—can also distort inference, especially when investigators do not report on statistical power. The concept of power is formally defined in terms of the random variability of results that is inherent in a specific combination of data structure, sample size, statistical models, and analytic method, but I believe that the concept should be substantially broadened to include the likelihood that a particular effect would be detected and reported if it were present to some specified degree. Such an analysis rarely accompanies "negative" findings, and readers may be left with an unjustified sense that an effect not demonstrated is an effect not present. Again, however, there are counterarguments: a report with low power may be better than no report (and no power), or meta-analysis[1] of several low-power reports may come to stronger conclusions than any one of them alone. Reporting negative studies of low power can create ethical problems, but those problems may be largely mitigated if the low power is accurately and clearly reported as well. Too many scientists resist the objective reporting of this kind of weakness in their work, and pressure for "strong" results may be greatest during the formative years of graduate training and career entry. Thus we may be training new scientists in unethical methods.

Despite the occasionally useful roles of these and other practices listed in Table 1, each can seriously distort the processes of inference and should therefore be an object of concern. Where the practices have legitimate applications, they should, of course, be used, but even then they should be fully and explicitly disclosed by the investigator, justified in some detail, and accepted with caution by readers. A combination of restraint in their overall use, limiting their use to clearly appropriate situations, providing full disclosure and justification, and maintaining the readers' skepticism will help to diminish the frequency and severity of ethical problems. Full disclosure here means more than a few words buried in the fine print of a Methods section; it means not just that the author sends a message, but that the author also works to ensure that the message is received and correctly interpreted by readers. There are parallels here to the evolving requirements for informed consent by experimental subjects.

Pressures to publish can be great and may account for many of the abuses suggested by Table 1. I fear that even the constrained use of these potentially damaging practices will leave attractive loopholes for an army of ambitious practitioners of science, each feeling great pressure to publish, who will rush in to explain why their situation is different, why full disclosure is inappropriate, and so forth. I am convinced that science, scientists, and society as a whole would benefit from substantially broader concepts, ultimately based on the need to protect the processes of inference, about ethical standards and violations in science.

Note

1. T. A. Louis, H. V. Fineberg, and F. Mosteller, "Findings for public health from meta-analysis," *Annual Review of Public Health* 6(1985):1–20.

More Squabbling Over Unbelievable Result

ROBERT POOL

Nature's publication of the investigation of an unbelievable experiment has triggered as much debate on the journal's conduct as on the truth or falseness of the results.

An investigation of unbelievable results that were published last month in *Nature* has raised more questions than it answered, particularly concerning why the results were published in the first place. The investigation, conducted by *Nature*'s editor, John Maddox, professional magician James Randi, and fraud investigator Walter Stewart, concluded there was "no substantial basis" for the claims put forth in the earlier *Nature* paper. In response, the scientist who did the research has criticized the conduct of the investigation and questioned *Nature*'s decision to accept the paper when the journal apparently suspected fraud or trickery.

In the 30 June issue of *Nature*, French chemist Jacques Benveniste and co-workers published the results of a series of experiments that seemed to have no physical explanation. The researchers measured the response of a type of human white blood cells to varying concentrations of a particular type of antibody. They diluted the antibodies with distilled water to the point where there should have been no antibody molecules left in the solution, and still they observed a reaction from the white blood cells. Standard theory offers no explanation for such a result, and the researchers suggested that the antibodies were somehow leaving an imprint on the water molecules that triggered the response of the white blood cells.

To convince *Nature* to accept the paper, Benveniste arranged for independent laboratories in Israel, Italy, and Canada to repeat the experiments, and researchers from these three labs were listed as coauthors on the final work. The journal held up publication of the paper for two years as it pushed for various substantiations, and finally published it with the condition that later an investigative team would watch Benveniste's group perform the experiments and file a report on the conduct of the work.

That report, which takes up four pages in the 28 July issue, damns Benveniste's experiments as "statistically ill-controlled, from which no substantial effort has been made to exclude systematic error, including observer bias, and whose interpretation has been clouded by the exclusion of measurements in conflict with the claims [of the researchers]." The investigating team depicts the experiment as one whose results were more likely due to the desires of the experimenters than to physical reality. The report suggests that the research team members, two of whom are doctors of homeopathy, wanted the experiments to succeed because that success would support some of the tenets of homeopathic medicine, which uses very small doses of various substances to cure ills.

Maddox said that Randi, who has made a name for himself uncovering trickery of various sorts, was included on the team because Maddox suspected some of the results might be due to fraud. "We thought it quite probable that there was someone in Benveniste's lab who was playing a trick on him," Maddox said. Randi found no evidence of conscious fraud, however, and Maddox said a more likely source of Benveniste's results was "autosuggestion"—one or more of the researchers seeing what they expected to see or wished to see.

Benveniste, replying to the report in the same issue of *Nature,* denounces the behavior and the conclusions of what he calls the "almighty anti-fraud and heterodoxy squad." He notes that neither Randi, Maddox, or Stewart has a background in immunology and claims that this ignorance caused various mistakes and misunderstandings in the investigation. More seriously, he charges that the investigation was more a witch hunt than a sober search for scientific truth. "This was nothing but a real scientific comedy, a parody of an investigation carried out by a magician and a scientific prosecutor working in the purest style of the witches of Salem or of McCarthyist or Soviet ideology," he told the French newspaper *Le Monde.*

The conclusions of the investigation and the controversy over how it was performed spotlight *Nature*'s original decision to publish the paper. Why, for instance, would a journal publish experimental results suspected of stemming from fraud or misinterpretation of data? Maddox said he was pushed to print the article because the French press had been alerted to the story and were spreading details of Benveniste's work.

Perhaps the more important question is: Why not wait four weeks and publish the paper at the same time as the report of the investigation? In 1972, when *Nature* published an earlier unbelievable result that turned out to be incorrect—that rats could be trained to avoid the dark by injection of the chemical scotophobin into their brains—the journal included a vigorous dissent by one of the paper's referees in the same issue. (A historical note: That referee was Stewart, and the scotophobin experiment was what got Stewart started on his crusade for accuracy in scientific publications.) Maddox said he decided not to publish the research article and the investigation together because he was concerned Benveniste would withdraw his paper upon seeing the report of the investigators.

Many scientists question *Nature*'s handling of Benveniste's paper. For instance, Arnold Relman, editor of the *New England Journal of Medicine,* said that, for such unbelievable results, the journal should have insisted on verification by a completely independent set of authors before publishing anything. What the journal should not have done, Relman said, was publish the paper and then undertake an investigation itself. "A journal should not be an investigative body," he said. An editor's job is to see that material is rigorously and fairly reviewed, he said, and when a journal acts as *Nature* did, "the editor becomes the judge, the jury, the plaintiff and—in some sense—the accused." Such a fraud investigation by the editor is a conflict of interest, Relman said.

The handling of the affair has certainly left a bitter taste in Benveniste's mouth. He says he would be happy to have someone point out errors in his experimental procedure that can account for the unbelievable result, but he feels betrayed by *Nature*'s decision to run the article and then attack it through the report of the three fraud investigators. "Everything has taken place as if one was trying to flush out a skylark from a field of corn in order to get a better shot at it," he said. "It may be that all of us are wrong in good faith," he wrote in his rebuttal to *Nature*'s investigation.

B: REPORTING AND FUNDING RESEARCH

I. DATA PRESENTATION

The issues presented in the transformation between the data obtained in the scientific laboratory and the presentation to the outside world of an interpretation of that data include the choice of forum for data presentation, the intellectual framework within which the data is to be presented, and the interactive process between the outside world, both scientific and societal, and the scientists presenting their findings. One fundamental distinction that must be drawn in this area is between written and oral communication.

A. Oral Communication

The opportunities for oral communication of a body of work occur on numerous occasions in scientific life. It can happen casually over lunch, in a chance encounter in the hallway, or in a planned, one-on-one discussion. Discussion of the scientific problem on which one is working in an informal manner is an essential part of scientific life.

However, oral presentation also occurs routinely under more formal circumstances. A scientist involved in a study that has not yet been published in written form may be invited to speak at another academic institution. This invitation provides the opportunity for airing of the data and its subsequent discussion in a forum that may provide valuable feedback and critical input. Researchers may choose to submit an abstract to a scientific meeting in their area. Either circumstance provides an opportunity to expose the finding to a large group of scientists who are knowledgeable and critical in the field in which the work has been carried out.

However, there are factors that restrain even the most informal types of communication. Of particular concern is the possibility that, by sharing scientific results at an early stage, a competitor may be able to carry out similar work and perhaps obtain scientific credit for it. This issue is not mere paranoia. A single critical result may open the door to a whole host of control experiments that should be performed by the careful scientist before formal exposure of a body of work to the scientific community. The possibility that another research group (with greater resources or less concerned with completing all relevant controls) may expropriate the research direction once the initial critical result has been introduced is a concern that can, and often does, restrict even informal scientific communication to those with whom a previous relationship of personal trust has been developed.

As the work proceeds and a point is reached at which the scientists involved in the study decide that the first formal presentation to the scientific world outside the lab is appropriate, various ethical issues arise. These include the degree to which one should be open in presenting ideas, planned experiments, and unclear findings and their interpretation, and also how to appropriately credit the contributions of others, and how to adequately present research results.

Increasingly, formal oral presentations are introduced with a slide listing co-workers and collaborators within and outside the lab who made significant contributions to the work presented, either intellectual or material. Frequently oral presentations are peppered with the names of individuals whose research has contributed to the theoretical framework and underlying assumptions of the work being presented. This is not only appropriate because it acknowledges the way research findings are interactive and build on each other, but it is also diplomatic if not politic since there are likely to be members of the audience involved in, or aware of, previous or ongoing related work.

Both oral and written presentation require the transformation of the data in notebooks into figures and tables that can be understood and evaluated by others. This process inevitably involves an abstraction of the data of some sort. Few, if any, formal scientific presentations involve a straight "data dump" from research records. But what to leave in and what to leave out? The rules of the game at this stage, in principle, are clear and straightforward. The data presented formally should be a clear and accurate representation of the results obtained. But, as we have noted earlier, this process inevitably involves selection of which data to present and in what form to present it. Does one, for example, present an early preliminary experiment that guided the choice of experimental design for a later critical experiment? Often the answer to this question is no. The guiding principle is that the findings eventually presented in a formal manner in a written format are presented so that they can, in principle, be reproduced by an independent experimenter. The experiments that represent "clues along the way" may not be reproducible even to the scientist in whose notebooks they reside because of the presence of initially unappreciated, uncontrolled variables. Much of the effort involved in bringing a finding from anecdotal observation to a form that can be accepted by the scientific community as a well-established observation may involve searching for and identifying the uncontrolled variables that underlie variation in the outcome of preliminary experiments. The systematic analysis of variation often is not presented in detail in formal scientific communication, despite its critical relationship to the final product.

Prior to submission to a peer-reviewed journal, oral presentation may involve a less stringent treatment of the data. Some oral presentation settings, for example, the Gordon research conferences, emphasize their informal nature. Written proceedings of such conferences are intentionally not kept, to encourage scientific communication of findings in a setting that comes closest to informal communication. Many larger scientific meetings involve submission of short abstracts. While data presented in these abstracts are not in any way considered to be peer reviewed, the abstracts nevertheless serve as a communication medium for scientists, which forces the presenters to consider what to include in their communication. Presentation at such a meeting often will involve a time limited oral presentation involving 10 or 15 minutes allocated to each speaker to present their findings. A complete rendition of the entire story, from concept to present state of affairs, is not possible in such a setting. Selecting graphics and photographs that quickly, clearly, and accurately convey the research findings being presented is not a simple task.

A presentation format that has become widely used in the past decade is the poster presentation. Because of its lack of permanence, and the fact that it is generally not peer reviewed, a poster presentation should be considered essentially a type of oral presentation. Yet it is not unheard of for posters to be photographed by individuals attending a poster session in order to have a lasting record of the presentation, and abstracts are sometimes cited in oral and written presentation. Because of the dynamic and semiformal nature of the verbal component of the poster session, it is a setting in which confusion, misunderstandings, and hyperbole may become a part of the information remembered by colleagues. For this reason, presenters should be especially clear about the message they present.

B. Written Presentation

No oral presentation format has the cachet equivalent to submission and acceptance of a paper by a peer-reviewed journal. It should be noted, however, that publication in a peer-reviewed journal does not make a finding "true." The fact that an article is accepted following peer review simply makes it likely that the underlying assumptions that go into the analysis of the data are shared by the authors of the article and the reviewers. This is not a guarantee that these assumptions are correct.

What peer review does attempt to guarantee is that the process of attempting to identify critical controls and analytical pitfalls has been carried out in a more formal and considered manner than is usually possible in the context of an oral presentation. A second function of the peer review process is to decide on the level of significance of the findings in the manuscript in relation to the mission of the journal. The submission of a manuscript to a journal therefore involves two sets of concerns on the part of the scientist involved in the study. First is whether the data and their interpretation are likely to pass muster from the perspective of the reviewer's view of scientific accuracy and experimental design. Second is whether the content of the manuscript is of sufficient import to fit the profile of the journal to which it has been submitted. To address the first issue the reviewer must be intimately familiar with the experimental techniques and approaches taken in the study. The reviewer is expected to point out flaws in experimental design and data interpretation. A common conflict that arises between submitter and reviewer is the complaint that "the reviewer does not properly understand the aspects of the paper that have been critiqued." The journal editor will attempt to avoid this issue by submitting the paper to at least two reviewers. If the two reviewers reach different conclusions about the paper, then a third or even fourth review may be requested. In most cases, the major issue of conflict relates to whether the manuscript will be published.

However, in some cases more troubling ethical issues arise. There is a variety of ways in which a reviewer may be in conflict in the review of a manuscript. Since the hope of the editor is that the reviewer will have a great deal of expertise in the scientific area covered by the manuscript, it is possible that the reviewer may in fact be working along parallel lines to the individual or group submitting the manuscript. A conflict of interest develops at the point at which the convergence of effort leads to a distortion in the judgment of the reviewer. Two remedies exist for this situation. One is the vigilance of the reviewer, who may be asked by the editor prior to receiving the manuscript, "Do you feel comfortable reviewing a manuscript with this title from these authors?" Second, the submitters, in their letter of submission to the editor, may explicitly identify one or two individuals with whom they are especially concerned regarding this issue. In general, an editor will honor requests to avoid a specific reviewer.

Even with these two safeguards built into the system, gray areas exist. If a reviewer has been seriously contemplating initiating a study prior to receipt of a paper for a review and the paper in some way confirms that the contemplated direction is a sensible one to take, is the reviewer now in a moral dilemma because information received in a protected form has a bearing on his or her own research direction? This dilemma is compounded by the fact that the reviewer and the submitter are bound not to contact each other: Their communication occurs strictly through the editor. This cast reflects the view that reviewers are more likely to be candid in their evaluation if they believe anonymity will protect them from retribution for a negative review. However, there is concern within the scientific community that some reviewers have taken unfair advantage of their anonymity to create obstacles for competitors or slow publication of a paper (by taking extra time to critique a paper, for example, or requesting additional experiments), in order to provide themselves with time and opportunity to get their own overlapping paper published first. One suggestion is that reviewers be identified so that there can be freer communication between authors and reviewers and so that authors themselves can monitor the related work of reviewers.

What if the reviewer realizes that the findings in the paper under review are relevant to an ongoing study of a graduate student such that when the paper under review is published, the assumptions that underlie the student's work will be proven false and the basis of that work vitiated? Again, this situation presents potential ethical dilemmas that exist within a framework that appears at the surface clear-cut and unequivocal: Until they are published, research findings that are under review are strictly confidential and the reviewer is bound to act in the remainder of his or her scientific life as though the paper under review did not exist. At the same time, members of the community have a responsibility to minimize wasted effort and resources, and to facilitate the career development of junior researchers.

The reviewer is thus caught between duties. The reviewer might wonder: If it is my own research, how could I *not* discontinue a fruitless or meaningless series of experiments once I am convinced it is a waste of time and effort? If it is my graduate student's thesis research, how could I not suggest an alternate strategy or approach? If it is another student on whose thesis committee I sit, should I not suggest an alternate way of looking at the problem? If it is a colleague, should I not ask pointed questions about the nature and validity of underlying assumptions? How much should I rely on the findings in the reviewed paper rather than repeat the experiments myself? The issues that may arise around the review process are not black and white because the scientific endeavor does not exist in a vacuum and the results reported in any one paper are inextricably intertwined with a larger whole.

C. Postpublication Ethical Concerns

Once the scientific paper has been published, ethical concerns do not cease. Initially, the publication may engender a demand from the scientific community to make available reagents or techniques described in the paper. This might be a specific strain of bacteria, virus, or other organism that can be propagated further in the recipient's laboratory. The ability to acquire the strain in question may be crucial to repetition of the work described in the paper. From the perspective of the basic tenets of the work, such a request is appropriate and should be responded to without hesitation. However, strategic and logistical questions come into play. The simple response to a very large number of requests for a strain can overwhelm the capacity of a small laboratory. A solution to this concern has been the creation of repositories that house and supply, at cost, strains of organisms that are of interest to the community.

A second consideration that may be relevant to the availability of such strains is the concern that the requesting laboratory will use it to carry out the very next stage of the study in which the originating laboratory is currently engaged. Concern about this issue can serve as an inhibition to timely publication of research results. In the absence of a clear social impact of the research in question, the choice as to pace of publication and subsequent availability of relevant reagents is more directly connected to the scientific style of the authors than to a moral imperative. Some individuals will prefer to explore a problem in greater depth prior to publication and release of reagents, while others are more anxious to bring the initial phase of the story into the purview of the entire scientific community. However, there may be additional pressures to publish when there is apparent competition, when the funder has requirements or expectations that research results will be published within a particular time frame, or when there is an apparent need for publication for the career advancement of the principal investigator or another member of the research group, especially a student, postdoc, or someone else looking for a job.

When research findings are expected to have an impact on the public, the importance of the rapid publication of research findings is increased. A tension develops between the rapid publication and the need to have a high level of confidence in the reliability of the data presented.

The transfer of a technique from one laboratory to another is a more complex matter than the transfer of a strain of a living organism. Often, it is not technically feasible in a printed form to truly teach the art required to master a newly developed experimental technique. While the authors have responsibility to transfer the technique they have developed to other research groups, the precise lengths to which the publishing group must go in order to fulfill this duty are not so clear-cut. If a technique is deemed to be of sufficient import to be broadly useful in the field, then workshops or courses to transfer the technique to laboratories unskilled in the technique are often developed.

What happens if contradictions arise to the findings and conclusions in the original study? At one level, the new findings may be part of a study of their own. In this case, a new publication that discusses the differences between the findings described in the initial publication and the present one may be the most appropriate approach to presenting the situation. Alternatively, the initial findings may have been determined to be in error due to a single source of error. An instrument may have been miscalibrated, for example. Under this circumstance the authors may choose to issue a retraction in which they state that specific conclusions reached in the original study are now considered to be invalid because of the specific source of error. This area includes vast gray zones. A short note of retraction may be the most direct way of dealing with an issue of this type, but may not address issues of significant scientific complexity to clearly convey to the reader the underlying technical issues. When a fuller explanation of technical matters is necessary, a longer paper may be required but needs to clearly identify itself as a retraction of earlier work. In general, scientific journals do not adequately provide for the need for this sort of retraction.

D. Presentation of Scientific Data and Findings to Society

The manner in which scientific findings are communicated to the public at large presents serious ethical issues for the scientist. The question of pace of publication takes on a new significance if the results have clear relevance to a significant societal concern. In principle, the choice of timing in presentation of such results to the public depends on two concerns that can be in potential conflict. One is to present results that may have important social implications in a timely fashion so that the results can be of use to both policy makers and the public. At the same time the responsibility to carry out the study to the point where the highest degree of accuracy possible has been achieved must also be considered.

Precisely how information on a scientific study is released to the general public is also a significant issue in this context. In general, public release of such data is appropriate at the moment of publication of a peer-reviewed article containing the data. Science reporters, while usually knowledgeable in general, are likely to have only a limited understanding of a new set of findings in a particular field. There is a responsibility for scientists to communicate to the public through the science journalist as clearly as possible. This means accurately portraying the limitations as well as the strengths of the study. One of the most difficult issues for scientists is the tendency for press reports to require a certain amount of hyperbole in order to reach publication. Scientists need to avoid becoming a part of an overstatement of results and/or their significance. This issue can be compounded by the desire of the scientist and the institution at which the work is performed to achieve notoriety in a form that may have a positive effect on the ability to raise future research funding.

In all of the venues in which the scientist interacts with the public, the duties of the scientist as responsible citizen as well as the duties of the scientist as designated expert add a level of depth and gravity to the role which the scientist plays in society.

II. AUTHORSHIP

Authorship is important for a number of reasons. Written papers are the primary way that researchers communicate their ideas and their experimental findings within and beyond the scientific community. Authorship is also the mechanism through which credit for one's efforts and contributions is allocated. Through authorship the scientific community measures scientific achievement and contributions "to the common fund of knowledge."[1] Publication is the coin of the realm, and authorship is analogous to patents for inventions and copyright for creative works of literature, art, music, and computer software. In some instances authorship may be linked to the determination of intellectual property rights assigned by these other mechanisms. It is also critical for professional development because one's publication record is the basis for hiring, and career advancement through promotion, tenure, and the awarding of funds for further research.

Because authorship is the primary mechanism for assigning credit for one's efforts, plagiarism, the usurpation of authorship by claiming another's written word as one's own, is considered unacceptable within the scientific community. Plagiarism is clear when an individual substitutes his or her name for that of the actual author, incorporates large portions of someone else's written work in an article without attribution, or identifies him- or herself as author of a translated article. Plagiarism can be more difficult to identify when the findings, conclusions, and recommendations of others are in transition into the realm of common knowledge and are paraphrased and incorporated into review articles. Regardless, plagiarism is considered a form of stealing.

While credit is one side of the importance of authorship, responsibility is the other. Authorship is the mechanism for assigning responsibility for the content of the work and all that it implies, including the ideas presented, the reliability and accuracy of the data presented, the accuracy of the description of methods employed, and so on.

One of the ongoing controversies associated with authorship is whether responsibility for the intellectual and factual content of a paper should be distributed equally among all authors, or allocated according to the particular aspects of the paper with which a researcher was involved.

Of continuing concern within the community of scientific researchers and publishers is the interplay between quality and quantity. To the extent that more publications can be obtained from a specific research study, an increase in numbers will lead to a longer resumé and seemingly greater productivity. Conceivably, the same information with minimal changes could be used in a research paper, a review article, an editorial comment, and a book chapter. This tends to promote and perpetuate the scramble for increasing the number of publications without an emphasis on quality. A similar issue is the habit of publishing small bits of distinct, but closely related, pieces of information that could be published together as a larger whole with greater effect. Each bit is described as a "Least Publishable Unit (LPU)"[2] and is analogous to a puzzle piece published separately rather than the puzzle as a whole.

A. Credit

In considering the allocation of credit associated with authorship, there are two issues to be considered: who should be an author and in what order the authors should be listed. In order to consider various standards of practice and common conventions, we need to identify the components of publishable research. Some divide the process into three parts: concept or problem definition; experimentation or observation, and calculations; and writing. Others subdivide the components further and identify experimental design, data collection, data analysis, and interpretation as separate segments of the process.

In general, authors are assumed to be all those who made a significant scientific contribution to the original, new information that is the core of the paper. It is assumed that it will be an intellectual contribution,[3] and will reflect active involvement in the design and execution of experiments.[4] Contention usually revolves around interpretation of the terms "significant," "original," and "new." These criteria would exclude various categories of contributors such as secretaries, programmers, statisticians, technicians, cartographers, funders, laboratory administrators, and providers of clones, reagents, viruses, and so on. This can be problematic since some who would be excluded, such as technicians, lab administrators, and providers of clones, have arguably been included as authors. Some organizations[5] explicitly include technicians and programmers among those whose contributions to the project potentially make them eligible for authorship.

Roger Croll,[6] in his landmark paper on the topic, emphasizes that hourly wages, academic credit, salary, and commission are irrelevant in assessing credit. This is counter to the rationale sometimes advanced as to why technicians, consultants, and others should not be listed as authors, that is, that they have been compensated for their contribution. This argument ignores the fact that generally all who contribute to the project are paid to do so, directly or indirectly. It also equates authorship with compensation.

Authorship is widely recognized as a reflection of *both* intellectual contribution and responsibility for the final product.[7] One view of the responsibility of authors is given by the Council of Biology Editors (CBE). In its 1983 *Style Manual,* the CBE argued that the basic requirement for authorship is that every author must be able to take public responsibility for the contents of the paper. Specifically, each author must be able to explain why and how observations were made, and how the conclusions follow from the data. It should be noted that it is the CBE's expectation that this would be the result of each author's meaningful participation at every stage, from experimental design through data collection, analysis, interpretation, and writing. These relatively stringent criteria would exclude even more individuals than the criteria of "significant . . . original, new information" described earlier.

An additional factor is the extent to which authorship is regarded as an important element in career development. Thus, for similar contributions of time, effort, and ideas, and a comparable degree of responsibility, a graduate student is more likely to be included as an author than is a technician. An exception to this is likely when promotions require authorship as they do for some technical positions.

In addition to authorship, there are other mechanisms for giving credit in scientific publications. It is widely agreed that those whose contribution is not sufficient to qualify them for authorship are appropriately recognized in the "Acknowledgments" section. Acknowledgments specify a particular, limited contribution to the larger work, usually more or less described in the acknowledgment itself. Similarly, intellectual debts for ideas or methodologies are recognized through a citation. Citations reference other work in the scientific literature that is the source of relevant ideas and information.

B. Order of Authorship

The order in which authors are listed has varying significance from one discipline to another, within the same discipline, and even among a single group of authors. There are no definitive rules regarding the order of authors.

1. Conventions

Most commonly in the life sciences, authors are listed in order of the nature and extent of their contribution. The first author may be the individual who had the original idea, who did the most work, or who wrote the first draft of the paper. In earlier times the senior scientist in whose laboratory or under whose auspices the work was carried out was listed first. More recently the senior scientist tends to be listed last, but sometimes first. In some fields, such as physics, and/or as required by some journals, authors are listed in alphabetical order. When a group of individuals publish more than one article together, the order of authors may rotate. If papers with a varying disciplinary focus are published from the same work (e.g., behavior and molecular biology), the author whose primary disciplinary orientation corresponds to that of the paper is likely to be listed as the first author. Should conflicts arise around the appropriate order of authorship when it is based on the extent of contribution, the senior scientist may choose an alternate convention, such as alphabetical or senior scientist first, to resolve the dispute. Within the scientific publishing community the perception is that "conventions of alphabetical or reverse-alphabetical schemes sometimes obscure the true 'first author'—that is, the person who takes the greatest responsibility and hence should claim the greatest share of the credit."[8]

2. Why Order Matters

Order of authorship can be significant because of the nature of commonly used reference styles in scientific publications. For articles with more than three authors, and usually with more than two, the reference to a source of an idea or technique is often indicated in the text by the last name of the first author followed by "et al." (for et alii meaning "and others") and the year. Thus reference to a paper by Smith, Jones, and Brown from 1994 will be cited as "Smith et al., 1994" in the text.

Thus, a reader will usually receive a concept linked with the name of the first author on the paper in which it is presented. Over time, a particular idea or finding is likely to become most commonly identified with the first author within, and perhaps beyond, the scientific community.

Order of authorship is also important because of assumptions on the part of the reader and the scientific community regarding the significance of order. Thus, if it is a commonly held notion that the first author is the individual who made the most important or greatest contribution to the work presented, then the order of authorship can have substantial implications for career development. For example, some institutions consider only those publications on which an individual is first author in making hiring or promotion decisions.

3. Why Order Doesn't Matter

At the same time, in an important sense, the order of authorship is not of overriding significance because of the Matthew effect.[9] The Matthew effect is named for the biblical passage from Matthew 13:12, "For unto every one that hath shall be given, and he shall have abundance: but from him that hath not shall be taken away even that which he hath." It refers to the observation that recognition accrues to those who already have the greater reputation, even when it is not deserved, and is at the expense of those who are lesser known but potentially more deserving because they made a greater contribution. Thus, a widely known and respected scientist will generally be considered the driving force behind an important idea or finding and will receive credit for an assumed contribution, regardless of a listing as first author, last author, or in the middle. This is not surprising since it is easier to remember something in connection with something (or someone) else that is familiar. Nonetheless, this tends to lessen the recognition of contributions of new, unestablished researchers.

However, the Matthew effect also provides advantages.[10] The reputation of the established researcher casts an aura of excellence, implying comparable quality and capability of those selected as students and colleagues. It is also suggested that an excellent researcher with proven capabilities brings out the best in others and facilitates focused thinking. In addition, a work associated with a well-known researcher may garner the attention of the research community to a topic or concept that might otherwise have been passed over. For these reasons, students and junior colleagues generally seek the imprimatur of credibility and quality that comes with co-authorship with a well-known scientist.

C. To Avoid Problems

It is always helpful to decide who will be authors before the paper is written. One approach is to circulate a list of all contributors indicating whether authorship, acknowledgment, reference, or no mention was suggested, and inviting additions or changes.[11] Ideally it is preferable to determine authorship even tentatively before the work is done.[12] In reality, this may be difficult because it is generally not possible to predict the change in direction, focus, and emphasis of a project as it develops. In addition, there is always the difficulty of distinguishing between original ideas, and the effect of the work, comments, and perceptions of others that lead to, or trigger, that idea.

For students, authorship can be both important and confusing. In some disciplines, students are increasingly expected to have a number of publications to their credit before they complete a graduate degree. The authorship convention and policy of a thesis advisor are usually not publicized, and the needs and expectations of a student can clash with expectations, styles, and professional needs of others in the research group. Students generally have the least power and the least experience regarding professional standards of practice.

D. Responsibilities of Authors

To quote Ian Jackson in *Honor in Science:* "The time to take responsibility for a paper is . . . before it is published." Each author is, at the very least, expected to have reviewed the manuscript, to have helped to refine and correct it, and to understand it well enough to discuss it. Similarly, each author has the corresponding right to have seen and read the manuscript in order to make contributions, refinements, and corrections that protect his or her professional reputation.

The CBE[13] asserts that each author has the responsibility to assure the community that proper ethical standards have been upheld in the conduct of the work presented. This includes assurance that the proper procedures were followed regarding the use of human subjects and animals such as consent forms, maintenance of confidentiality, approval of research protocols by an Institutional Review Board (IRB) or Institutional Animal Care and Use Committee (IACUC), maintenance of professional standards regarding pain control, and so on. In addition, the CBE guidelines indicate that each author has the responsibility to make sure that the paper cites all related work that substantiates or that contradicts the work presented in the paper, and that the paper includes only and all observations actually made. These expectations regarding the responsibilities of authorship have added complexity in the context of collaborative research.

E. Collaboration

Some of the most interesting and exciting research at the forefront of science involves collaboration between investigators from different disciplines. Differences in expertise are the basis for collaboration. But researchers in one specialty are likely to have a limited understanding of the details of the science involved in another specialty: the scientific theories that form the basis for the techniques employed and the rationale for problem solving in the field, the pitfalls of instrumentation and measurement. When it comes to assuring the accuracy of the work involved, an element of trust is required.[14] However, trust is not enough. A certain degree of discreet investigation into the reputation of potential collaborators is appropriate if a senior scientist is to be able to assure students and junior members of the research group that he or she heads that the collaboration is sound and will not endanger developing professional reputations. It has been proposed that others in the field of the collaborator be called on to review the work independently.[15] This may be a valuable source of insight, but may prove difficult to accomplish and raises ethical problems of its own if the discipline is small or highly competitive. In other words, competition in a discipline limits the extent of openness with which investigators may be comfortable.

III. THE PROCESS OF OBTAINING FUNDING

Scientific research costs money. Salaries of investigators must be paid; equipment and supplies must be purchased; physical space to carry out the research must be acquired and the costs of its maintenance must be covered; services ranging from the washing of glassware to time on computers must be obtained. The need to acquire funds to meet these costs can raise ethical issues for the scientist.

Funds to cover these costs are derived from a number of sources. Research grants from government organizations are a major source of funding. Grants from private nonprofit organizations, donations from private individuals, and funds from private industry also make up a significant part of this mix. Two basic forms of research support are the research grant and the research contract. The research grant proposal is concerned with the justification of the significance of the questions that the scientist proposes to test, the description of the research strategies and techniques that will be used to test the hypotheses, and the description of the methods of data analysis that will be used. One key distinction between the research grant and the research contract is the responsibility for hypothesis generation. The scientist applying for a research grant is responsible for the generation of hypotheses both in the grant application and in the modification of these hypotheses in response to experimental findings during the course of the research. In the case of a research contract, the hypothesis to be tested, as well as the line of investigation to be pursued in terms of method and strategy, is explicitly laid out by the organization providing the funding. A research contract often will have specific milestones tied to the achievement of explicitly stated research goals. Deviation from the original line of investigation is usually not an option in the performance of a research contract.

The ethical issues that arise during the research funding process bear much resemblance to the ethical concerns that arise in the peer review process. Information provided confidentially by grant applicants may be relevant to ongoing work by reviewers. As in the peer review process for publication, the grant reviewer must deal with information provided for grant review in a manner that is separate from the remainder of the reviewer's scientific life.

Decisions regarding funding have many implications. Such decisions can have a substantial impact on society as well as a major impact on the career development of the scientist applying for funding. Often societal considerations are implicit in the grant funding process. For example, depending on societal priorities, some areas of research are preferentially supported over others (e.g., AIDS or breast cancer research).

A scientist may be faced with ethical dilemmas when he or she is committed to the type of research being funded, but disagrees with the motivation behind the funding. Consider, for example, the extent of funding for research on communication with marine mammals supported by the U.S. Navy. Researchers might strongly agree with the importance of funding research in this area, but strongly disagree with the possible use of the information for military purposes.

The funding process also influences career development and stature within the scientific community. The ability to garner research funding is often a key element in professional status. An ideal of fairness is thus essential in the development and implementation of research grant application evaluation. However, just as in the peer review process for publication, there is no simple set of rules that can ensure "fairness," since the process involves a combination of assessment of scientific merit (in and of itself a somewhat subjective criteria) and relevance to the program goals of the funding entity.

A. RESEARCH GRANTS

The research grant programs of the major governmental agencies, the National Institutes of Health, National Science Foundation, Department of Energy, Department of Defense, and others, operate under a system of peer review. A two-tiered system of review is used to determine how these funds are awarded. The peer review system involves a detailed study of each research grant proposal that is submitted to the agency by a study group composed of scientists who are established investigators in the field of research germane to the proposals under review. The mission of this group is to review the proposals submitted with respect to scientific merit. This rating process is intended to be carried out independent of "programmatic" concerns. In the second tier of review, a second, independent group, which includes both scientists and individuals representing a broader spectrum of concerns from a societal perspective, uses the scientific merit ratings from the initial round of review to establish a prioritized ranking of the proposals. At this stage the proposal may receive an opportunity for higher priority for funding due to its substantial relevance to the mission of the funding body.

The issue of scientific merit versus program relevance encapsulates some of the inherent conflicts and issues in the process of evaluating grant proposals for funding. The motivation to provide funding for research in a particular area is the consequence of goals both specific and general that society has chosen to express through the allocation of resources to the funding agency. These goals may address a specific human problem (e.g., a cure for a spinal cord injury), in terms of a broader scientific mandate (e.g., an understanding of the elementary particles of physics), or even a foreign policy issue, as occurred following the launch of the first earth satellite in 1957 by the Soviet Union. When the mission of the agency or enabling legislation is stated in narrow terms, significant conflicts between what is the "best research" and which research is "most likely to lead to a cure for AIDS" inevitably occurs. While the funding agency must resolve this dilemma on a continuing basis, which research directions will ultimately lead to the desired result are not always easy to sort out. A strong argument can be made that the most relevant research for a particular goal may, in fact, come from a broader, less programmatically oriented research program. For example, the most significant contributions to understanding the underlying basis of breast cancer may come from work on cell division in yeast.

The strategy of private, nonprofit agencies in awarding research grants is often even more programmatically guided than government funding. Although a single review body may serve as arbiter of both scientific merit and relevance to the goals of the organization, relevance to the mission of the agency is often an even more critical concern among private agencies.

Apart from the concerns related to program mission, significant concerns about the fairness of the review process inevitably occur. As in peer review for journals, grant review almost always involves two or more reviewers who critique the grant in depth, as well as a broader discussion by a group of 10 to 20 individual reviewers. In general, when the review is carried out by a government agency, a written report that summarizes this review process is made available to the applicant soon after the review is completed. Private organizations working with smaller budgets and staff levels often do not have the resources to provide equivalent written reports of the outcome of grant review proceedings to the applicant. While all of this provides no guarantee of fairness, this system of review has a significant number of opportunities to ferret out reviews that are prejudicial or simply not competent. Granting agencies and even small private organizations make a significant effort to respond to concerns from applicants regarding the fairness of a grant review. Nevertheless, since the fund/no-fund decision acts on the very lifeblood of the research enterprise, decisions in this area are among the most central and sensitive in science, and the potential for significant harm by unethical conduct in this area is self-evident.

Once the grant is awarded, the grantee is expected to carry out the research program outlined in the proposal. However, as the research program develops, new directions, unanticipated in the original proposal, often become apparent. In accepting a research grant, the researcher has committed to carrying out the best possible research program he or she is able to carry out in the area of investigation outlined in the proposal. As new information is gathered both by the researcher and others in the field, it is essential to modify the research design. The grantee's responsibility is to weigh these new inputs against the original intent of the grant and to adjust priorities and scientific strategies and goals.

B. Research Contracts

Award of a research contract from a government or private sector source has a different set of parameters and consequently a different set of social and ethical concerns compared to the awarding of a research grant. A research contract generally is developed as a consequence of the identification of scientific goals, by the agency or company providing the funding, that can be best met by a scientist or scientific group outside the agency or company carrying out the work in question. The social and ethical issues that arise in the performance of work under such a contract are often the consequence of control over the broad direction of research being removed from the hands of the scientist who receives the funding. What happens if research performed under contract reveals an unexpected side effect for a newly synthesized chemical compound? The company sponsoring this research may legitimately wish to cut its losses and suspend further research on this compound. Yet a scientist involved in such a study might be motivated, and indeed feel compelled, to explore in depth the mechanism underlying this side effect. This is precisely the type of unanticipated research direction that a scientist would be expected to pursue under a research *grant* but that would not be supported by a research *contract.*

The issues that surround the need to garner and maintain financial support for research persist throughout the active research life of a scientist. The dynamic tension between this necessity and all of the other concerns to which a scientist must respond can cause many moments of soul searching during a scientific lifetime.

Notes

1. R. K. Merton, "The Matthew effect in science: The reward and communication systems of science are considered," *Science* 159(1968):56–63.

2. W. J. Broad, "The publishing game: getting more for less," *Science* 211(1981):1137–39.

3. R. P. Croll, "The noncontributing author: An issue of credit and responsibility," *Perspectives in Biology and Medicine* 27(3)(1984):401–7.

4. R. A. Day, "How to list the authors," In *How to Write and Publish a Scientific Paper,* third edition (Phoenix: Oryx Press, 1988), 21–27.

5. Martin Marietta Document Preparation Guide Section 3.1.4.3: Guidelines regarding criteria for authorship. May 1986.

6. Croll, "The noncontributing author."

7. CBE Style Manual Committee, "Ethical conduct in authorship and publication," *CBE Style Manual: Guide for Authors, Editors and Publishers in the Biological Sciences,* fifth edition (Bethesda, Md.: Council of Biology Editors, 1983) 1–6); Croll, "The noncontributing author"; Day, "How to list"; C. M. LaFollette, *Stealing into Print: Fraud, Plagiarism, and Misconduct in Scientific Publishing* (Berkeley: University of California Press, 1992).

8. LaFollette, *Stealing into Print.*

9. Merton, "The Matthew effect."

10. Merton, "The Matthew effect."

11. Martin Marietta, Guidelines.

12. Day, "How to list."

13. CBE, "Ethical conduct."

14. D. Baltimore, "Baltimore's travels," *Issues in Science and Technology* 5(4) (1989):48–54.

15. I. Jackson, *Honor in Science,* second edition (Research Triangle Park, N.C.: Sigma Xi, 1991).

CASES FOR CONSIDERATION

CASE 1

Dr. Gorchek has asked a senior graduate student to look over a grant proposal that Gorchek has submitted to the USDA. In reading the proposal the student recognizes passages taken directly out of an invited review article that she and Gorchek published in the Annual Reviews of Immunology, but the passage is not referenced. Another long passage that looks familiar, the student learns, was reproduced exactly from an annual progress report sent to a different funding agency in connection with another grant awarded last year. The student is concerned and discusses that concern with a postdoc in the lab, who assures the student that plagiarism only occurs when one fails to reference or acknowledge someone else's writing.

Questions for Discussion

1. Is the postdoc correct that plagiarism only occurs when someone else's writing is not referenced?
2. Does it matter that the material that was copied is found in a grant proposal rather than in a published paper?
3. What would you recommend that the senior graduate student do at this point?

CASE 2[1]

You have an interesting idea, but pursuing that idea will also require the technical skills of another colleague's lab. The colleague is intrigued and you pursue the project together; the work also involves your postdoc (A), a postdoc in the colleague's lab (B), and the colleague's technician. When all is said and done, people have made a variety of contributions. You had the idea, paid for the portion of the work done in your lab, and regularly discussed the work with your postdoc. Postdoc A did virtually all the benchwork from your lab, constituting 60% of the data. Your colleague developed the idea in the area of her expertise, paid for the portion of the work done in her lab (which was significantly more expensive than yours), and supervised her people. Postdoc B spent several hours per week on benchwork, but the technician did the bulk of the data collection coming from the colleague's lab. Now it's time to write the paper.

QUESTIONS FOR DISCUSSION

1. What role should each of these investigators take in writing the paper?
2. What are the functions of authorship?
3. What factors should be included in deciding who should be an author? In deciding the order of authorship?

Note

1. This case was developed by Catherine P. Cramer, Department of Psychology, Dartmouth College.

Pressure to Publish and Fraud in Science

PATRICIA K. WOOLF

Q. Is there pressure to publish in today's science?

A. Yes, there is.

Q. Is pressure to publish bad for science?

A. Bad or good, we're stuck with it. Since its birth, modern science has had an ethos that called for communications; scientists have felt some pressure to make their results known. Publication is an established way of doing so.

Q. Should there be pressure to publish?

A. Why not? If the public has paid for research, they are entitled to evidence that research has been done.

Q. Are publications the product of research? Doesn't the public pay for new knowledge and ideas?

A. Yes, that's the ideal, but not all experiments work. The public is really paying for a good-faith effort by highly skilled people who collectively have a pretty good chance of coming up with some information that is an advance.

Q. How do we know which scientific results represent advances in scientific knowledge?

A. That's where publication comes in; information known only to the discoverer cannot be evaluated and become knowledge. Some say it isn't even science until it is published. Research reports are evaluated by editors and referees before they are published, and then publication allows other scientists to further evaluate and use the results.

Q. Why is there so much concern about pressure to publish?

Reprinted with permission from *Annals of Internal Medicine* 104(1986):99–102.

A. Many persons in science are worried that their important but complicated system of communication is in jeopardy. Because of undue pressure it may not be working the way it's supposed to.

Q. Why is it complicated?

A. Publication is no longer just a way to communicate information. It has come to be a way of evaluating scientists; in many cases it is the primary factor in professional advancement.

Q. Is that so bad? If articles are refereed, doesn't that ensure a breadth of professional judgment?

A. Some say that the quantity of published papers has become more important for evaluation than the quality of ideas they contain.

Q. Is that true?

A. It's hard to say. The evidence is anecdotal. Promotion and granting committees meet in secret; according to a dean in a major medical school, "Confidentiality fans the flames of paranoia." There are no systematic studies of these deliberations, but some reputable people have criticized the system. We need more systematic information about the relation of publications to professional advancement.

Q. Why are they so critical if the facts are not known?

A. Their experience has led them to be concerned about science. And they are worried that careerism is damaging a very important national resource—our research establishment.

Q. Is there that much pressure to publish?

A. It depends on what sort of pressure you mean.

Q. Is there more than one kind of pressure to publish?

A. There are at least two sorts: the pressure the scientist feels from the tradition of science, and the pressures from institutions.

Q. Are these pressures the same?

A. How could they be? Institutions and their departments may have different standards for promotion. Requirements are often implicit rather than explicit, and frequently they are misunderstood. Some research fields publish more heavily than others.

Q. Why have so many of the recently disclosed frauds been found in medical schools?

A. No one knows for sure. Their research environments are highly competitive. The heroic nature of their quest directly affects issues of life and death, so there are greater economic and social rewards in medical science. Another possibility is that some fraud occurs in all science, but it is more likely to be detected in a rapidly growing research area like biomedicine.

Q. Is there more pressure on medical scientists than on others?

A. Some say there is. Medical researchers often have clinical responsibilities in addition to teaching, research, writing, and applying for research funds.

Q. Have any studies measured psychological pressures on scientists?

A. I haven't found any in a search of the literature. One study currently underway at a major medical school is trying to find out if persons who are up for promotion really understand the promotion and tenure procedures of the institution.

Q. How can we measure the amount of pressure being put on scientists by institutions?

A. One way is to look at the amount of publishing done by practicing research scientists in our best institutions.

Q. Aren't you just counting papers again?

A. It's not the best way of determining research outcomes, but if becoming a full professor at a good university takes an exorbitant amount of publishing, we probably have a problem on our hands.

Q. How many papers does the average scientist publish?

A. A study by King and colleagues[1] showed that in 1977 the average number of articles published per physical scientist was 0.20/year, up from 0.17/year in 1965. The average number per life scientist in 1977 was 0.40/year, up from 0.30 in 1965. In this period only the social sciences showed a big decline; the average number of papers per social scientist was 0.22/year, down from 0.60 in 1965. But these data include industrial or government scientists who may not be actively publishing. A better definition of scientist is probably a "publishing scientist." Price and Gursey[2] have determined that among publishing scientists, the average productivity is 1 paper per scientist per year.

Q. What are the rates of publication of scientific faculty in our major universities?

A. They differ among disciplines. In a study of the faculty of nine distinguished universities, rates ranged from 1.8 publications per scientist per year in physics to a high of 2.7 publications per scientist per year in biochemistry. Astronomy, chemistry, and microbiology were the other specialties studied. Publication records of over 1000 faculty members were examined.[3]

Q. What about medical schools?

A. There are many medical schools whose faculty publish very little. In the 1973–1975 period the average rate of publication for privately controlled medical schools was 250 papers per year and for publicly controlled medical schools, 150 papers per year.[4]

Q. What do federal granting agencies expect from their investments in research? Is there unreasonable pressure to produce simply to satisfy these requirements? On the average, how many papers per grant per year are produced?

A. The Office of Program Planning and Evaluation of the National Institutes of Health (NIH) has examined the literature output of NIH-supported research. A recent study of faculty in departments of medicine shows that one third of M.D. faculty are not significantly engaged in research at all.[5] The NIH makes between 5000 and 6000 competitive grants per year. The average length of the grants is 3.2 years and that amounts to roughly 16,000 to 20,000 grants in force. In the same period and with the predicted lag of 3 years between a grant award and its resultant publication, 16,000 to 20,000 papers per year are produced. These figures are subject to a possible 20% to 30% increase, because at different periods some journals were not included in the survey. But that amounts to 1 or 1.3 papers per grant year.[6]

Q. If you double that to 2 or even 3 papers a year per grant, would that be too much pressure on a scientist?

A. It doesn't seem like it.

Q. If these institutional expectations are relatively modest, why is fraud associated with pressure to publish?

A. There are two principal reasons. First, several of the most spectacular frauds have taken place in laboratories where the number of papers published every year significantly exceeded the norm. Second, many of the persons involved have mentioned pressure as a factor. The newspapers usually report that view, and pressure has come to be a stock explanation.

Q. Could we look at those two factors separately?

A. That's sensible. "Pressure" has come to be a convenient explanation, for two reasons. The reports of these events have shown how shocked scientists are to discover that one of their colleagues was not playing by the rules. They may explain that fault using a psychological excuse that focuses on the flawed personality of an individual person and not on science in general; it is a less severe condemnatory judgment than, for instance, moral turpitude.

Q. Is that what's called medicalizing the deviance?

A. Yes, it seems to isolate the problem and let the other scientists off the hook. But then questions began to be asked about where the pressure comes from.

Q. Does that bring us round to the publishing practices in laboratories where fraud was discovered?

A. Let's look at just a few. The first of the major recent scandals, that of William Summerlin at the Memorial Sloan-Kettering Cancer Center, was in the research group supervised by Robert A. Good. In the 5 years before 1975, Good had published 342 papers, an average of 68/year (16 as first author, 325 as co-author with at least 136 co-authors). Several other frauds were associated with medical research groups that have been extremely prolific. In the period 1975 to 1980, which preceded the discovery of fraud in their laboratories, Eugene Braunwald had published 171 papers, an average of 29/year. Philip Felig had published 191 papers, an average of 31/year.

Ephraim Racker, who is not on a medical faculty, had published an average of 16/year.[7]

Q. How do these pockets of prolific publication compare with similar departments in similar institutions?

A. These scientists have made important contributions, significant far beyond the numbers of their publications. Therefore, comparisons with average productivity have limited value. But, yes, their average levels of publishing are significantly higher.

Q. Should we be worried about such prolific publishing?

A. There is reason for concern on several levels. Simple time-and-motion studies lead to the conclusion that when an author's productivity is extremely high, his participation in each paper has to be low. Or at least there is no clear way of telling which articles the author contributed to significantly. These prolific scientists are clearly operating in ways very different from those in the mainstream of science. One can argue that excellent science has always been out of the mainstream, but when there are significant deleterious effects on younger co-workers in their laboratories and eventual effects on scientific research in the United States as a whole, we have a responsibility to question the wisdom of those research and publishing practices.

Notes

1. D. W. King, D. D. McDonald, and N. K. Roderer, *Scientific Journals in the United States* (Stroudsburg, Pa.: Hutchinson Ross, 1981), 62.

2. D. Price, and S. Gursey, "Studies in scientometrics: Part I. Transience and continuance in scientific authorship," in *International Forum on Information and Documentation* (Moscow: International Federation for Documentation, 1975), 17–24.

3. J. D. Frame, "Quantitative indicators for evaluation of basic research programs/projects," *IEEE Transactions in Engineering Management* 30(1983):106–12.

4. P. R. McAllister and F. Narin, "Characterization of the research papers of U.S. medical schools," *Journal of the American Society for Information Science* 34(1983):123–31.

5. H. N. Beaty, D. Babbon, E. Higgins, P. Jolly, and G. S. Levey, "Research activities of faculty in academic departments of medicine," *Annals of Internal Medicine* 104(1986):90–97.

6. Analysis of research publications supported by NIH and NIGMS, in *NIH Program Evaluation Report* (Bethesda, Md.: National Institutes of Health, 1980). Also, H. Gee, personal communication.

7. *Science Citation Index Source Index* (Philadelphia: Institute for Scientific Information, 1970–1980).

Guidelines on Authorship

INTERNATIONAL COMMITTEE OF MEDICAL JOURNAL EDITORS

At its last meeting the International Committee of Medical Editors (the Vancouver group) drew up the following guidelines on authorship and on other contributions that should be acknowledged. The committee also expanded the section on information to be given in the covering letter to include details of any conflict of interest and clarify the position of the author responsible for final approval of proofs. These guidelines will be incorporated into the Uniform Requirements for the Submission of Manuscripts to Biomedical Journals[1] when it is next revised.

GUIDELINES ON AUTHORSHIP

Each author should have participated sufficiently in the work to take public responsibility for the content. This participation must include: (a) conception or design, or analysis and interpretation of data, or both; (b) drafting the article or revising it for critically important intellectual content; and (c) final approval of the version to be published. Participation solely in the collection of data does not justify authorship.

All elements of an article (a, b, and c above) critical to its main conclusions must be attributable to at least one author.

A paper with corporate (collective) authorship must specify the key persons responsible for the article; others contributing to the work should be recognized separately (see Acknowledgments and other information).

Editors may require authors to justify the assignment of authorship.

Reprinted with permission from the *British Medical Journal* 291 (1985):722.

ACKNOWLEDGMENTS OF CONTRIBUTIONS THAT FALL SHORT OF AUTHORSHIP

At an appropriate place in the article (title page, footnote, or appendix to the text; see journal's requirements), one or more statements should specify: (a) contributions that need acknowledging but do not justify authorship, (b) acknowledgments of technical help, (c) acknowledgments of financial and material support, and (d) financial relationships that may constitute a conflict of interest.

Persons who have contributed intellectually to the paper but whose contribution does not justify authorship may be named and their contribution described—for example, "advice," "critical review of study proposal," "data collection," "participation in clinical trial." Such persons must have given their permission to be named.

Technical help should be acknowledged in a separate paragraph from the contributions above.

Financial or material support from any source must be specified. If a paper is accepted it may also be appropriate to include mention of other financial relationships that raise a conflict of interest, but initially these should be outlined in the covering letter.

INFORMATION TO BE INCLUDED IN THE COVERING LETTER

Manuscripts must be accompanied by a covering letter. This must include: (a) information on prior or duplicate publication or submission elsewhere of any part of the work; (b) a statement of financial or other relationships that might lead to a conflict of interests; (c) a statement that the manuscript has been read and approved by all authors; and (d) the name, address, and telephone number of the corresponding author, who is responsible for communicating with the other authors about revisions and final approval of the proofs.

The manuscript must be accompanied by copies of any permissions to reproduce published material, to use illustrations of identifiable persons, or to name persons for their contributions.

Note

1. International Committee of Medical Journal Editors, "Uniform requirements for manuscripts submitted to biomedical journals," *British Medical Journal* 284(1962):1766–70.

Rustum Roy: PR Is a Better System Than Peer Review

IVAN AMATO

What would you do if you were a materials scientist in academia who had come up with an innovation that just might put your university in a position to profit from the growing synthetic diamond industry, already a $500 million-a-year business? Submit a paper to *Science* or *Nature,* then wait for peer review to recognize your ingenuity? Not if you're Rustum Roy; the much-published, controversial, knife-tongued founder of Pennsylvania State University's Materials Research Laboratory (MRL) has criticized researchers who hype their results and journalists who buy their claims uncritically. But last week Roy and his colleagues played a few media games of their own, putting out a press release and hiring a room in Washington's National Press Club to publicize what they think is a breakthrough in synthesizing diamond more easily, efficiently, and in a wider variety of shapes and sizes than can be done with conventional methods.

Not that Roy and his two MRL research associates completely eschewed the ordinary process of publishing scientific results. They submitted separate, though related, manuscripts to *Science* and *Nature.* But Roy held the press conference and distributed the manuscripts to reporters long before the editors at the two journals had a chance to review them—a practice widely known to reduce a researcher's chances of publication in the elite scientific press.

But Roy doesn't think that's such a huge loss. He claims peer review is a terrible process: too slow and too "leaky," allowing peer reviewers to

Reprinted with permission from *Science* 258(1992):763.

gain research advantages unfairly. "I have absolutely no faith in peer review," he told the assembled press. Roy told the press conference he chose the mass media as a way of getting his results out because they're faster than peer review—and because they're a more effective way of drawing the attention of corporate executives. Although Roy says, "I'm against hyping things," he claims his behavior in this case doesn't constitute hype, because his discovery is the real thing. This is "no cold fusion," he said.

Roy's research peers express puzzlement about his end run around peer review. Noting that Roy's research record is sufficient to secure him plenty of respect, one researcher, who insisted on anonymity, said: "I honestly don't know why he is doing this. This is an interesting idea but he should have just sent it off for publication"—and peer review. William Banholzer of General Electric, a leader in the world's synthetic diamond industry and expert at PR to boot, added that the Penn State group "may have wanted to make as big a splash as they can because it might attract students and funding agencies."

Easy to lose in the bemusement over Roy's unorthodox ways is what could be a new way of making synthetic diamond. Other synthetic diamond researchers who have seen the manuscripts consider the work interesting and potentially important. In its simplest variant, the Penn State researchers combine carbon-rich powders with nickel powder (which acts as a catalyst for diamond formation), spread the mix atop a crystalline surface that can serve as a growth template for diamond, then shower the preparation with hydrogen plasma (known to encourage diamond growth at the expense of other carbon structures) while baking it all afternoon in an oven at about 1000°C. During the process, atoms in the inexpensive carbon powders rearrange into diamond's more pristine and valuable geometry. The technique should be able to transfer "virtually any form of solid carbon" into diamond of arbitrary shapes and sizes, the researchers claim.

This process would certainly seem to be a promising alternative to the most commonly used industrial processes for making synthetic diamond, which require vast pressures and temperatures and produce only small diamond particles, used mostly for cutting tools. And most cutting-edge research on synthetic diamond making focuses on converting gaseous (not solid) carbon sources, such as methane, into thin films of diamond using a family of techniques called chemical vapor deposition (CVD). The use of a solid carbon source in a low-pressure technique, and the option of making larger, thicker diamond objects, renders the technique "radically innovative," Roy claims.

But other researchers aren't persuaded that Roy's technique is all that new. Crystallume, a Menlo Park, California, firm, has several patents for techniques that produce "diamond ceramics" by subjecting a preshaped aggregate of diamond particles to CVD conditions. Even Roy's more basic speculation that the Penn State synthesis might be occurring by a novel solid-to-solid transformation, which has never been seen at the low pressures of the technique, has its doubters. J. Michael Pinneo of Crystallume and others suspect instead that carbon from the solid is probably first vaporizing into gaseous fragments and then redepositing a diamond onto the solid's surface. That would make it "a variant on a lot of previous [CVD] research," says John Angus of Case Western Reserve University, one of the field's most respected practitioners. Still, "if he can convert large pieces of porous graphite into diamond, this would be positive," Angus adds.

The kinds of expert judgments Angus and Pinneo were offering are precisely what constitutes peer review. In considering what inspired the Penn State diamond makers to short-circuit that process, clues might be found in market analyses that project synthetic diamond to become a multibillion dollar business by the end of this decade, a fact Roy pointed out in the press conference. The MRL, which coordinates a multicompany diamond research consortium, is well aware of the financial stakes—and the zeal of their worldwide competition. Hence the PR-for-peer-review switch, which Roy and MRL director Russel Messier say was blessed by the university's provost and by patent attorneys. "We [at universities] have not been efficient at converting research into patented, protected technologies," says Messier. "This [PR campaign] is forging new policy at Penn State for protecting important results."

MRL's self promotion has generated results: The *Wall Street Journal* wrote a piece based on Roy's press release, as did several magazines including *Science* and *Chemical & Engineering News*. Only time and scrutiny by other researchers will decide if the science Roy so eagerly publicized last week will prove as innovative as his PR tactics.

Peer Review: Treacherous Servant, Disastrous Master

CHARLES W. McCUTCHEN

Peer review, out of control, makes science a jungle where politics rules and fraud is tolerated.

Science has become a profession: grants and research contracts are what it lies on. Whereas a rich dilettante like Lord Rayleigh could retire to his country estate and do acoustics or whatever else he wanted, modern scientists must sing for their supper. They do not sing to their patron, the U.S. tax payer. They sing to other scientists, who wield over them the power of professional life and death via peer review.

Peer review, the evaluation of a specialist's work by others in the same field, is an inevitable consequence of specialization. Example: though anyone can tell if a bridge design is truly bad—the bridge collapses—it makes sense to have other engineers check the plans before the bridge is built. Science uses peer review to determine which projects to pay for and which articles to publish, and, recently, to judge cases of alleged misconduct.

Peer review suggests trial by a jury of one's peers, a jewel in the crown of Western democracy—surely an excellent model. But it takes more than a jury to have a fair trial. A lynch mob is also a peer panel. Rules and procedures—jury selection, rules of evidence, the requirement that evidence be heard in public—and a judge to interpret and enforce them are necessary if fallible people are to render fair decisions. Specialist

Reprinted with permission from *Technology Review* (October 1991):27–40, copyright 1991.

peer review is fraught with biasing influences. Specialists compete with one another and, at the same time, fight collectively for their profession.

Peer review is at best a treacherous servant, but scientists often forget that a jury trial is more than a jury, and act as if the use of peers automatically sanctifies the resulting decisions. Establishment scientists have been treated well by peer review; scientific administrators use it. Both want to believe in it, and the need engenders beatification by faith. "Peer review is the distinguishing characteristic of science," they say. "It makes science what it is."

They are right—in a way. Every scientist is an informal peer reviewer. A scientist's work affects science only if others accept it. But formal review of grant applications, manuscripts, and fraud allegations also makes science what it is, and here human failings can yield improper decisions whose practical consequences and poor ethics propagate throughout science.

Peer review resists investigation. Only insiders know the details of each decision. They may not tell the truth, and the technical background needed to extract the facts is hard for outsiders to learn. Lacking the omniscience of Orwell's Big Brother, we must be content with horror stories of reviewing gone wrong. Though such stories do not directly reveal the frequency of mistakes, they show which human failings are involved, and thus the likelihood of trouble and how to reduce it.

PEER REVIEW AND GRANT GIVING

The federal government uses a variety of ways to decide how to fund science. Department of Defense (DOD) managers can fund whoever they like, without having to get advice. They do not compete for contracts with the scientists they might choose. Instead, they shine in the success of the programs they manage, and should something go wrong in a program, the manager is responsible. These are all good features. Unfortunately, managers are subject to agency politics.

As consultant to a small firm, I watched the Navy fail to give a fair hearing to our best idea, *Sea Knife,* a fast boat of strange but simple shape that rides smoothly in rough water. We decided that the Navy's small-boat people would not admit that a craft by outsiders might be better than theirs. But having figured a way around this obstruction, we were funded to build *Wavestrider,* a faster though rougher-riding and more complicated boat. We got unrelated contracts to explore far-out and ultimately unsuccessful forms of marine propulsion. I think these were funded because they threatened no powerful group within the Navy.

At the National Science Foundation (NSF), too, managers make the final funding decisions but with the advice of peer reviewers. Managers benefit from the peers' specialized knowledge but have the authority to correct for peer bias. As at the Defense Department, should something go wrong, the program manager is responsible. For years, George Koo Lea, director of the fluid mechanics program, supported Van Chao Shein Mow, now of Columbia University, for work that was never novel and true at the same time. Workers in his field, the lubrication of animal joints, who disagreed with the professor had trouble getting funded by NSF.

At the National Institutes of Health (NIH), where I have worked since 1964 in biomechanics, optics, and fluid mechanics, peer reviewers effectively make the final decisions; managers are nearly powerless. In each discipline, a peer panel—the study section—evaluates grant applications. By secret ballot, each panel member gives an application a numerical score, and these scores largely decide its fate. An upper, advisory council can fund projects slightly out of the order of their scores without attracting comment, as can program managers. But when whistleblower Robert Sprague, a grantee for many years, did well in the study section but lost out at the advisory council, the event made news.

Since peer review puts a scientist's future at the mercy of competitors, is it any wonder that career issues are a respected, if unadmitted, influence on decisions? Would we not expect mutual assistance pacts to be accepted facts of life? Should we be surprised that politics is especially rife in disciplines funded by NIH, where the power of scientists over one another is essentially unchecked? Van Mow receives three-fourths of all NIH support of research in joint lubrication and still accomplishes little. For years, those with contrary views received none. Support for research on lubricin, the lubricating chemical in joints, ceased in 1982.

Since power over grants confers power elsewhere, dissent in joint-lubrication research appears only in unrefereed publications such as conference reports and public lectures or in journals of distant fields. Timothy Harrigan and his then supervisor, MIT biomedical engineer Robert Mann, made an important contribution to the theory of how cartilage in joints deforms. Refused by the *Journal of Biomechanics,* it was accepted by *Archives of Rational Mechanics.*

If NIH grant administrators made the final funding decisions, they could be called to account for permitting cronyism. But peers are the ultimate authority, and because they exercise this power in secret, no one is accountable. This unaccountability makes the NIH system attractive to management. When a Professor Mow seems to have an inside track, NIH blandly declares that his success shows that his colleagues think well of him. Whoever gets funded, NIH can say the decision was out of its hands. However deserving Dr. X from Rep. Y's district may be, administrators can say they have no way of influencing X's funding. Privately, NIH officials admit there is politics in study sections but say it is a price worth paying to insulate grants in biomedical science from national politics. The cost was surrendering control of funding to scientific politicians.

NIH has one potential lever. The executive secretaries of study sections, who are NIH employees, appoint section members and could use that power to suppress the politics. Although members, who serve for four years, cannot succeed themselves, they expect a large say in the choice of their successors. A section secretary could threaten, "If you misuse your power, your successor will not be from your faction." But such action would require support from NIH management, because section secretaries are not famous scientists. The support would not be given, since NIH conceals this power of appointment. The handbook describing the study sections says that their members are "selected by the NIH" but supplies no details. The impression given is that peer review is above the vulgar mechanics of the appointing process. Without support from above, an executive secretary would need great courage to stop a determined cabal from controlling NIH funding in a discipline.

So cronyism proceeds. In 1976, Mow and Peter Torzilli published two spectacularly erroneous papers on joint lubrication. NIH knew experts held the papers to be nonsense. They predicted such a rapid flow of fluid through the porous joint cartilages that viscous losses would have generated heat 100 million times faster than mechanical work was being done on the joint. Yet Torzilli replaced Mow when the latter left the Orthopedics and Musculoskeletal Diseases Study Section in 1984.

INHERENT FAILINGS OF THE PROJECT GRANT SYSTEM

Politics is particularly bad in biomedical research because biomedical scientists directly control the flow of money that supports their disciplines. But even without politics, today's grant system, in which scientists propose future research projects to an agency, would be bad. The system gained its popularity after World War II when there were fewer scientists and most projects were funded. But even in those flush times, the Office of Naval Research, Atomic Energy Commission, and National Science Foundation all refused Donald Glaser when he asked for funds to develop the bubble chamber, later the standard device for observing particles in high-energy physics.

The great ideas in science in the next few years will be those not yet thought of. The system ought to select people likely to think them, but, alas, it is inherently biased against such speculation. Granting agencies want certainty, and reviewing peers fear unexpected discoveries by their competitors. As NIH puts it in a pamphlet for grant applicants, "Reviewers prefer limited clear goals that can be realistically approached; rather than broad, multiple questions or vague goals the attainment of which is open to doubt."

The caution of officialdom displays itself in a 1940 report from the Gas Turbine Committee of the National Academy of Sciences: "Even considering the improvements possible . . . the gas turbine could hardly be considered a feasible application to airplanes mainly due to the difficulty in complying with the stringent weight requirements." Thus did great men, including engineer fluid dynamicist Theodore von Karman, evaluate the turboprop and jet engine.

The project grant system ignores the range of human talents. As well as inflicting anguish on inventive people, it has no official niche for promoters, people who make enterprises go, people like Vladimir Zorykin who persuaded RCA to spend its money developing television. The grant system does not eliminate such promotership; it just perverts it. Promoters pretend to be great and impossibly active scientists to get money in promoter-scale quantities. They sign every manuscript from their laboratories and accumulate reputations earned by the work of others.

The grant system disrupts the chains of authority and loyalty between scientists and universities. Each university scientist is like a pirate ship raiding the U.S. Treasury. The university provides docking space; in return, the scientist provides for his or her keep, and preferably more, out of grants. To the scientist, the university is a leech; to the university, the scientist is a prima donna. In the middle of these cross purposes, students are supposed to be taught.

PEER REVIEW OF JOURNAL ARTICLES

When peers referee journal articles, they perform a valuable service. They find mistakes and sometimes fraud, and they form a trial readership whose reactions show what to change to hold a reader's attention. A referee who knows the field can clarify what is and is not novel in a manuscript. Competent reviews take hours or days of hard work and are a tribute to those who do them.

Unfortunately, the power of referees, usually anonymous, permits self-interest, jealousy, revenge, and other unworthy motives to influence decisions. Dozens, probably hundreds, of letters to the editor over the years show that nastiness in reviewing contributes to a general unpleasantness in the publication process and in science as a whole.

Reviewing weeds out good manuscripts as well as poor ones. Frederick Lanchester's 1894 circulation theory of how wings lift, Chandra Bose's photon statistics in 1924, Enrico Fermi's theory of beta decay in 1933, Herman Almquist's discovery of vitamin K_2 in 1935, Hans Krebs' citric acid cycle in 1937, and Raymond Lindeman's trophic-dynamic concept in ecology in 1941 all were turned down at least once. Charles Fourier and Gregor Mendel had trouble getting published. We will never know how many deserving manuscripts remained unpublished.

The time and energy spent fighting to be published are lost forever. Lindeman died before his article appeared, and the delay Almquist suffered may have cost him a share in a Nobel Prize. The discoveries by Fermi, Almquist, Krebs, and Lindeman were held up only for a short time, but the circulation theory of lift was delayed over a decade.

My experience has been similar. Since Lord Rayleigh's time, it has been known that the wave nature of light spreads the image of a point source into a blur whose shape on the focal plane is described by the two-dimensional Fourier transform of the lens aperture. The image projected by a square lens is a diamond-shape array of checkerboard squares. I realized that a Fourier transform relation between aperture and image also holds true in three dimensions. When I attempted to publish this fact in the *Journal of the Optical Society of America*, referees rejected it. The editor kindly published it in response to my plea. This relation is now the starting point for calculating the three-dimensional resolving power of confocal scanning microscopes.

This evidence is anecdotal, so, by current convention, those who find it uncomfortable can ignore it. But in 1977 Michael Gordon wrote in the *New Scientist* that Henry G. Small of the Institute of Scientific Information had found "a significant negative correlation between referees' evaluations of [highly cited chemistry] papers and the number of citations the papers subsequently received." Low citation scores followed high opinions by referees, and vice versa.

The inability of peer reviewers to judge good papers should be no surprise. A discovery is usually a better-than-his-or-her-average product of a brighter-than-average (or perhaps simply unusual) scientist; the resulting paper will likely be reviewed by an ordinary scientist, operating at an ordinary degree of inspiration, and possessing human imperfections. Truly novel papers may not be understood. Those understood will be envied and perhaps rejected with one excuse or another. In 1844, J. J. Yaterston tried to publish a paper that anticipated by several years the kinetic theory of Clerk Maxell and Ludwig Boltzmann. A referee pronounced it "nothing but nonsense, unfit even for reading before the [Royal] society."

These famous examples of rejected discoveries end with World War II. With the rise of grant-supported science, few manuscripts are unembellished reports of discoveries. A discovery is too valuable to reveal in a journal article until it has been used in grant applications. By the time most discoveries are published, they are already on the rumor circuit, and the papers announcing them include data generated in work the grants paid for.

It is follow-up papers that most scientists write and that referees are most likely to approve. A paper starkly describing something new looks strange and will be treated like the proverbial ugly duckling. An example: theoretical treatments of a plate planing on the surface of water like a surfboard demand that a sheet of fluid be ejected forward from under the plate. I found this not so in practice. Instead, there is a tumbling mass of foam where the plate meets the water. When I tried to report this in the *Journal of Fluid Mechanics*, none of the four referees disputed my findings, and three complimented my work. But the four were unanimous that my manuscript could not be published in the journal. One said my account was too sketchy even for a grant application. So far as I know, official fluid dynamics has not yet acknowledged the phenomenon, though my article is now Appendix D in *Planing* by Peter Payne.

Publication can lead to jobs and research support: NIH hired me as a result of my publications on joint lubrication. By denying publication to unadorned discoveries, refereeing obstructs this career channel and drives innovators to the granting agencies and ultimately to the establishment. A deadening uniformity is enforced. Dilettantes are squeezed out, not because they are bad scientists but because they do not belong to the union. This is a major loss. A Parisian gardener was the first to reinforce concrete with steel. Lanchester, inventor of the circulation theory of lift, was a mechanical engineer, not a fluid dynamicist. The inventor of the traveling-wave amplifier was trained as an architect, and two musicians invented Kodachrome.

Adding to the number of scientists by drawing from the fat middle of the bell curve of ability may retard rather than accelerate progress. As reviewing peers, the new recruits may silence and starve better scientists out of science. This happened to Douglas Kenyon, who once calculated the flow of water though joint cartilage. He now works for the Marathon Oil Co., and calculates the flow of petroleum through rock.

I call the cooperation of referees with the establishment an "evolved conspiracy." Referees, doing what their personal devils make them do, force innovators into the arms of the establishment, and the establishment is happy with this fact. Were it unhappy, changes would be made.

MISUSING PRIVILEGED INFORMATION

Reviewing of journal articles and grant applications gives reviewers the intellectual pleasure of interacting with authors and proposers, as well as education that, I suspect, has led to more advances than generally realized. These rewards are legitimate. Some rewards are not.

An obvious misuse of privileged information is rejecting or delaying a competitor's paper. The anonymity of referees ordinarily renders this untraceable. In 1978 Vijay Soman and Philip Felig rejected an article on anorexia by Helena Wachslicht-Rodbard and others to ensure priority for an article of their own. The action was detected only because the offenders plagiarized the reviewed article, and their manuscript was sent to Wachslicht-Rodbard for review.

Under cover of anonymity, reviewers can steal ideas from grant applications and manuscripts. There have been many private complaints by apparent victims. Theft is hard to prove, but it is known that the composition of the first material that was superconducting at the temperature of liquid nitrogen was leaked from a paper that Maw-Kuen Wu et al. submitted to *Physical Review Letters;* the leak was revealed because yttrium was wrongly called ytterbium in the manuscript. This error turned up on the grapevine.

A few proved cases do not show that stealing is common. But the rewards are large, especially now that professors must win grants to get tenure and promotions. It is bad form for victims to complain in public. Indeed, it is half-accepted that big fish will appropriate the success of little fish. Jocelyn Bell's discovery of pulsars won a Nobel prize for her superiors but not herself. There was an outcry but not of the size the injustice deserved, nor did the superiors seem embarrassed.

PEER REVIEW AND FRAUD

The current attempt to deal with scientific fraud is science's first brush with formal self-regulation. Self-regulation of any profession runs afoul of collective self interest and pack loyalty. When disciplinary committees operate in secret, these influences have full rein. Need I enlarge on the ineffectiveness of the disciplining of doctors by doctors?

Though a few fraud cases are famous, most investigations have been ineffective: a top NIH administrator told me that no university can bring itself to use the word "misconduct." He exaggerated. A very few small fry have been found guilty—for example, the unfortunate Lonnie Mitchell of Coppin State College in Baltimore. He had his grant application prepared by a professional writer who plagiarized someone else's application that Mitchell had provided as a model. Alas, the plagiarizee reviewed Mitchell's proposal.

The vast majority of scientists who stand accused before a university bar of justice are exonerated. Tim Beardsley recently reported in *Scientific American* that the accused was found guilty in only 16 of 110 cases completed by the Office of Scientific Integrity (OSI) since it took over as NIH's fraud squad in early 1989. According to Lyle Bivens, head of the Office of Scientific Integrity Review (OSIR), which oversees OSI for the Department of Health and Human Services (DHHS), NIH has reversed only one university exoneration. At face value, this says that most fraud charges are baseless, but we have only the word of the universities and OSI that this is true. Details of the exonerations, including the names of accuser and accused, are secret. (I am suing DHHS under the Freedom of Information Act in an attempt to lift this secrecy.) Where secrecy has been penetrated, exonerations have been found to be mistaken. Both the University of Wisconsin and OSI declared James Abbs innocent of Steven Barlow's charge that he had forged an illustration for a journal article by making a smoothed tracing of a figure in an article co-authored by Abbs and Barlow. *Neurology* has published a letter to the editor in which I demonstrated the relationship between the figures. Abbs' published response gave no satisfactory explanation of the resemblance.

A little-known case is revealing. The University of Medicine and Dentistry of New Jersey, and later OSI and OSIR, all told Gene L. Trupin that he was wrong in claiming that Barbara Fadem had stolen his research. OSI and OSIR ignored obvious signs of trouble. Just one example: in defending herself and other members of the university faculty against a lawsuit by Trupin, Fadem said that a journal article he and she co-authored proved that Trupin knew certain facts when the article was submitted. Court records show the facts in question were added to the article at the proof stage, 10 months after the date of submission. OSI knew about this dodge at the time it found Fadem innocent. It also knew that the suit was settled out of court in 1988 with a $60,000 payment to Trupin.

As long as NIH's watchdog is blind to evil when it wants to be, is it any wonder scientists learn that ethical pliability is a professional necessity, and find it prudent to discover that what looks like fraud is a "scientific disagreement," an "error," or "sloppiness"?

One might think a determined whistleblower could force OSI to conduct a real investigation. Not so. Once OSI receives an accusation, it tells the whistleblower little or nothing. As the whistleblower who got the Abbs case reopened, I was volunteered no information: OSI's predecessor office did tell me to prepare a 10-minute presentation, but I was never summoned to make the presentation, nor told it was called off. Meanwhile, OSI's impenetrable secrecy encouraged Abbs to complain that his constitutional rights to due process were being trampled. He sued and won on a technicality. DHHS is both appealing the verdict and, as the judge required, going through the steps laid out by the federal Administrative Procedures Act.

Universities routinely use peer panels to investigate and judge fraud. This shifts responsibility but does not get justice done. A powerful accused scientist or pack solidarity can frighten a panel into seeing no evil. The panel that the University of Wisconsin convened to investigate Abbs' alleged faking ignored blatant inconsistencies in his submission. For example, Abbs falsely claimed that accuser Steven Barlow had displaced one record before comparing it with the other. The public gaze might shame a panel out of doing a whitewash, but panels operate in secret. Incredibly, in its filing under the Administrative Procedures Act, DHHS proposes that determinations of guilt no longer be printed in the *Federal Register.* Secrecy, secrecy, ever more secrecy.

Secrecy gives full rein to subterranean forces, and a major scientist can bring great force to bear. Panels at MIT, Tufts, and NIH all said, wrongly, that no misconduct was involved in a paper co-authored by Thereza Imanishi-Kari, Nobel laureate David Baltimore, and others. It is a matter of record that Baltimore used both a letter-writing campaign and professional lobbyists in an unsuccessful attempt to get Congress to halt Rep. John Dingell's (D-Mich.) investigation of the matter. (It was Rep. Dingell's investigation that finally forced NIH to mount a real investigation of its own.)

Media interest in the Baltimore affair is more than instinctive celebrity chasing. Fake work impedes progress much more if a major scientist is involved than otherwise, because others must pretend to agree with it if they want jobs or grants. I know of no attempt by other scientists to duplicate the precise experiments in the Baltimore affair. Scientists supposedly delight in proving one another wrong, but they hesitate to embarrass someone with power and the willingness to use it.

Because no one at NIH is accountable for the decision to fund Professor X, no one feels betrayed, no one is angry or ashamed if X commits fraud. So NIH washes its hands of the matter and passes off the consequent cover-up as political realism. As an official in the Department of Health and Human Services said to me about OSI: "They have to compromise." Expedient exonerations are excused as being for the good of science. If the public got the idea that a lot of fraud exists, the argument goes, it might not support research. The whistleblower is, figuratively, given a loaded pistol and told to do the proper thing.

According to the *New York Times,* retired Harvard microbiologist Bernard Davis believes it would have been better had the Baltimore affair been dropped. The biomedical science establishment would rather let fraud continue than have it publicized, a policy that will keep fraud going forever. Concealment requires that the sinners keep their funding. Abbs and his laboratory received millions in government support after the initial brushing off of the complaint against him. So long as such scientists are protected and fed, their species will multiply.

By not using its control over who gets funded, NIH has given up the power that would go with being pay master. Despite signing 57 billion a year in checks for research, NIH was unwilling or unable to prevent MIT's whitewash of Imanishi-Kari, Wisconsin's of Abbs, and numerous similar instances.

Were NIH to invoke its power of the purse, a university might say it was applying improper influence, a confrontation NIH evidently fears. James Wyngaarden, ex-director of NIH, and Joseph E. Rall, ex-deputy director for intramural research, have both said that universities have run ineffective investigations, but NIH has never punished—or even tongue-lashed—them for doing so. Nor has it said that running a bogus investigation is unethical. Yet unless NIH greatly expands OSI, the agency will depend on university investigations of fraud.

Compare NIH with NSF, where managers make the final decision about who gets funded. With responsibility comes accountability—for such odd decisions as siting the National Earthquake Engineering Research Center in Buffalo, N.Y., rather than in California. One can also question the reasons for moving the National Magnet Laboratory from MIT to Florida State University. But whatever one may think of them, these decisions show that NSF has power. If NSF wanted a university to investigate a fraud, the school would remember the movability of laboratories before doing a whitewash. Perhaps this power is reflected in the apparent lack of fraud in the parts of science NSF funds.

TAMING A FRACTIOUS HORSE

• *Reform the Grant System.* Suppose politics could be eliminated from NIH study sections. Suppose DOD and NSF program managers were all smart and incorruptible. The project grant system would still be a time-destroying Moloch, demanding and reviewing long applications, most of which are not funded, and it would still sponsor sure things rather than imagination. Block grants to universities would be better. The schools would decide who to support however they wished, using any system they wished, from despotism to democracy. Universities have made good choices in the past. The University of Michigan found the initial, essential money for Donald Glaser's bubble-chamber research.

Each year, universities would go to the federal government and argue for support. Let them bring citation scores, rumors of Nobel Prizes almost awarded, whatever they want. Out of this free-for-all, a formula would emerge, no doubt with loopholes and exceptions, and the negotiators would return home exhausted and tell the troops how they made out. The mutual dependence of scientists and brass would develop the loyalty upward and downward that makes institutions bearable to their members.

Under the block-grant system, everybody in a university would be in the same lifeboat and would benefit collectively from one another's success. Still, researchers would continue to compete within the school, so to dull the teeth of university politics, perhaps 10 percent of federal support should remain as grants to individuals.

• *No-fault Publication.* Specialist journals should never reject. If scientists are worth paying, they are worth hearing from. A referee who thought a paper wrong could try to argue the author out of publishing it, invoke a six-month cooling-off period, impose a length limit of a page or two, and have signed comments published along with the paper. If no-fault publication results in a flood of garbage, it shows that scientists are creating garbage. Better we learn about this than conceal it.

General-circulation journals like *Science* and *Nature* would still reject most manuscripts they receive. Their editors, not reviewers, should make the final decision. Editors are the filter that catches reviewer misbehavior. Essay-form reviews can be windows into a reviewers motives, and having one reviewer from outside the specialty under review is a wise precaution against discipline politics.

Editors of all journals should ask reviewers to be as kind as possible, and authors should know the identity of writers of adverse reviews. A referee whose identity is known is less likely to steal from a paper, reject or delay it for professional advantage, or be pointlessly nasty. On the other hand, favorable reviews should be anonymous to discourage reviewers from trying to curry favor with authors. There is no way to keep them from informing authors privately, but the rule would remind them it is unethical.

- *Fraud.* The fraud problem reflects the ethics at the top of biomedical science. By not retaining for itself final authority over funding decisions, NIH left this power unguarded for ambitious scientists to pick up. With power came arrogance and the feeling that rules were for lesser beings. The cure is obvious. End the carving of their own cake by biomedical scientists, and the steamy politics will dry up.

If funding is not reformed, the scientific establishment will remain the problem, and the solution must come from elsewhere. John Dingell cannot interest himself in every fraud case, so the public's sense of fair play must be enlisted as a force for justice. Whistleblower and accused should know everything that occurs at every stage of an investigation so they can object and, if necessary, complain in public. The final conclusions of all fraud investigations should be made public.

If a peer panel has to make the final decision, as it might in cases of fraud, only extraordinary measures will yield justice. Because panel members are specialists judging fellow specialists, precautions beyond those in jury trials are needed to counter the effects of politics and pack loyalty. Accused and accuser, or their advocates, must have the right to question panel members in public about decisions before they become final. Unless these or very similar reforms are instituted, OSI should be closed, because it cannot yield justice.

Using peer review is like riding a fractious horse. One must understand its bad habits and never let it forget who is boss. Kept under control, peer review can yield good advice. Given its head, it will hurt people, serve the interests of the reviewing peers, and warp the institutions that use it. Where possible, peers should not make the final decisions but should advise the decision makers, who can filter peer self-interest from peers' recommendations. As a fractious horse is only as good as its rider, peer review is only as good as the program managers and editors who use it, but these people are visible and can be called to account for their decisions.

6

Conflicts of Interest and Conflicts of Commitment[1]

PATRICIA WERHANE and JEFFREY DOERING

I. INTRODUCTION

One popular view of the scientific method postulates that scientists should study nature in a value-free manner without any biases.[2] Despite the impossibility of its complete realization, this functions as an ideal for scientific research. Scientists, like all other people, have a variety of interests and commitments of an intellectual, personal, or financial nature that could conflict with a purely unbiased approach to their work, and compromise good scientific judgment. Recognizing and finding ethical ways to deal with these conflicts is the subject of this chapter. Take the following example:

Dr. ST, an M.D. from National Taiwan University with a Ph.D. from the University of California at San Francisco, became interested in the use of a vitamin A ointment as a remedy for keratoconjunctivitis sicca, an eye disease that prevents tearing, when he was studying with Dr. AEM at Johns Hopkins University. ST first tested vitamin A on rabbits under a series of federal research grants that AEM and he received from the National Institutes of Health (NIH). Having apparently achieved some temporary success with rabbits, he began testing on human subjects, first at Johns Hopkins and later at Massachusetts Eye and Ear Infirmary, a prestigious clinic that is linked to the Harvard Medical School, where ST received a two-year fellowship. At Massachusetts Eye and Ear the initial studies of the drug were carried out with the permission of the hospital's Human Studies Committee, which, under federal guidelines, evaluates all experiments that involve human subjects. This committee approved initial studies of 25 to 50 patients. As a result of these studies ST published a paper stating that "all patients demonstrated clinical improvements in symptoms."[3] At the same time AEM and ST set up a company, Spectra Pharmaceutical Services, to manufacture the vitamin A ointment. Shares in Spectra were publicly sold, and AEM and ST became the majority shareholders.[4]

According to press reports on this story, in fact, ST tested many hundreds of patients without permission of the Human Studies Committee. He subsequently produced two favorable reports on the ointment and began to use the ointment, an ointment now manufactured by Spectra, on patients at Massachusetts Eye and Ear. According to the reports, however, his test results had not proven whether or not the ointment produced positive long-term results. A later study performed by researchers who were not shareholders of Spectra in fact resulted in an unfavorable report about the effects of the ointment. Moreover, according to testimony of a number of infirmary nurses, ST often tested other drugs that had not been approved on patients' eyes, often without their knowledge or approval. No patient's eyes were injured, but the vitamin A ointment has proved to be ineffective as a long-term solution to "dry eye."

ST has since left Harvard and is a physician at the University of Miami. The *Boston Globe* investigation concluded that he and his family made at least $1 million on the sales of the ointment and his Spectra stock. Because of the publicity surrounding the ST affair, both the president of Massachusetts Eye and Ear and one of their leading researchers, Dr. KK, have left the infirmary. Harvard investigators later concluded that ST's involvement with Spectra violated university rules on conflict of interest. Eventually charges brought by the State of Massachusetts Medical Board were dropped against KK and ST. The Board decided that KK had not been party to ST's activities. Charges against ST were dismissed, despite the fact that he was found to have violated the infirmary's protocol policies, because of his other contributions to medicine and because there was no long-term harm to patients.[5]

This case illustrates the difficult issue of conflict of interest in scientific research, and it shows what can happen when conflicts are not recognized, made public, and remedied. ST had financial interests and professional aspirations that may have conflicted with the norms of good scientific research practice and human subject experimentation procedures. ST wanted to find a remedy for "dry eye," like all scientists he sought professional recognition, and like most of us, he desired financial rewards for his discoveries. But he failed to test his drug thoroughly, appeared not to have reported those test results that were negative, and he tested the drug without permission of the Human Studies Committee. These are all violations of acceptable scientific procedure. Moreover, he apparently tested drugs other than the vitamin A ointment, for which he had no approval and without patient knowledge—a violation of the norm of informed consent. ST had a conflict among his financial interests, his personal desire to be well known in his field, and the unbiased judgment expected of him as a scientist. Such conflicts are to be expected.

However, it appears that ST acted without paying attention to his conflict of interest. That is, he continued his work without informing others of his conflict. Had he done so, he might have been able to seek independent confirmation of his results. As a consequence of his disregard for his conflicts, he was less than thorough in following acceptable scientific research procedures. There may have been, in fact, no improper motives, but failing to recognize his conflict of interest, or acting in disregard to it, created a situation that threatened his scientific judgment.

Despite the ideal, it has never been true that scientists simply observe the world unencumbered by theory or bias. One can never "observe nature without prejudice." Earlier in this century the physicist Werner Heisenberg noticed that the act of observing subatomic particles affects the activities of those particles. He concluded that one cannot completely separate the observer from the observed. Without observation there is no object, yet the very act of observing affects what is being examined. This basic characteristic of some kinds of physical observations serves as a reminder of what is also true of the scientific process in general. Whether scientist or not, each of us perceives and experiences the world from a set of perspectives and theories that focus, schematize, and organize what and how we perceive. This "theory-laden" nature of science and scientific investigation means that pure objectivity is impossible. Nevertheless, the notion of value-free science or purely objective research provides the basis for identifying theoretical and other biases.

Even with the difficulty of achieving pure objectivity, and even though personal commitment is important to being a scientist, one can distinguish personal, professional, and scientific interests. How one makes these distinctions needs some explanation. Scientists, like all people, have multiple interests and goals—to be good friends, spouses, or parents, to be collegial, to be successful in their work, to make a decent living, as well as to achieve success as teachers, researchers, administrators, and/or writers. Thus, each scientist occupies a number of roles. A role is a position "defined by a set of rules or practices that indicate what that person should do [in a certain set of contexts]."[6] Roles refer to "repeatable patterns of social relations . . . which are structured partly by the rules of acceptable behavior."[7] Role responsibilities are spelled out by institutional, social, and/or cultural expectations of how a person should behave in that role. Roles are impersonal, that is, they define a position, not a person. An individual scientist, then, may also occupy the roles of spouse, parent, research associate, professor, member of a professional association, university administrator, citizen, and so on. Each of these roles has defined expectations, and in each role the person is held accountable in terms of these expectations.

Because of their complexity and diversity, some roles compete or conflict with one another. For example, having a financial interest in the company producing a drug you are studying can create a clear conflict, particularly when the studies are not entirely successful. Competing with colleagues for funding or wanting to be "first" to make a discovery may create conflicts for an individual who is reviewing grant proposals or manuscripts. Being both a parent and a researcher can involve conflicting role commitments because of competing demands on one's time. Passionate commitment to a particular theory can conflict with careful interpretation of new experimental data. This is sometimes the biggest conflict for a scientist, since there is an ego investment in how he or she wants the results of a study to come out. The existence of conflicting interests and commitments, then, is not uncommon for a socialized, passionate, interested human being functioning in a complex world. For the scientist the challenge is to recognize the existence of various interests and commitments and keep inappropriate ones from threatening or becoming the controlling aspect of scientific judgment. The challenge for the scientific community is to create public policies and guidelines that will help individual researchers avoid conflicts.

ST apparently acted without regard to the conflict between his financial interests and professional duties. But scientists and engineers have a number of conflicts. One way to sort out conflicts is to distinguish conflicts of interest from conflicts of commitment, a common distinction outlined in a number of policies for conducting research.[8] Conflicts of interest are those that exist between professional interests and personal or financial interests.[9] What distinguishes conflict-of-interest situations is that the conflict is between what one is trusted or expected to do in one's role (with its duties) as a scientist or engineer, and financial or personal influences or interests that will or could compromise one's professional judgment in that role.

Conflict situations involving an individual's financial stake in the outcome of a research project tend to be emphasized, but the investigator's professional stake (recognition) in a research result can be at least as critical. Conflicts of commitment are conflicts that entail a conflict between two or more sets of professional commitments that will affect one's focus of time, attention, and responsibility. For example, a well-known professor may have obligations to her profession to lead and give symposia, but also has an obligation to her university to teach and mentor graduate students. Because of the magnitude of her responsibilities, she finds it difficult to adequately honor both commitments.

II. CONFLICTS OF INTEREST

Having conflicting interests and commitments is part of the human condition. Only sometimes does a situation arise that may provide an individual with the temptation to compromise professional judgment for financial or personal gain. This is a conflict-of-interest situation. Where possible, researchers should avoid conflict-of-interest situations since they provide the temptations to act unethically.

The word "conflict" implies that the interests at issue do not coincide. Conflict-of-interest situations are ordinarily those in which all interests may not, or in some cases, cannot, be realized simultaneously, and where choosing a financial or personal interest over a professional one may violate a code or norm, a promise or contract, or some other specific professional responsibility. "conflict-of-interest" describes a situation that has the possibility for irresponsibility or poor judgment, which can potentially produce negative outcomes for others or oneself. But "[the existence of] a conflict of interest by itself does not indicate wrongdoing—it merely refers to a setting in which factors exist that might influence one's conduct."[10] One can consider such circumstances as a continuum, where competing interests can lead to conflict-of-interest situations that, in turn, can develop into unethical behavior. In some cases the distinctions can be somewhat arbitrary, but clear delineations will facilitate the discussion.

The existence of conflicts of interest will be distinguished from actually acting without regard to that conflict. In the latter cases, when one acts in disregard to a conflict-of-interest one may succumb to placing financial or personal interests ahead of those of science or engineering. In these cases one's professional role as a scientist is compromised. Moreover, in these cases compromised judgment can negatively affect scientific results so that harm to some persons, some institutions, or to the advancement of science itself is usually a result.[11]

Defining scientific conflicts of interest from a moral point of view,

"conflict of interest" refers to any conflict between research or other professional scientific judgments and financial or personal interests where *acting* with disregard to that conflict by placing one's personal or financial interests ahead of professional interests compromises or detrimentally influences professional judgment in conducting or reporting research.[12]

Conflicts of interest raise ethical issues because (a) the integrity of the scientist or engineer as a professional researcher may come into question, (b) trust in what one expects of the scientist as a professional researcher may be betrayed, (c) scientific research may be unduly biased in favor of personal or financial interests, and/or (d) there can be harmful outcomes.[13]

According to the press reports in ST's case, (a) he did not follow traditionally specified scientific procedures involving human subjects, thus bringing into question his role as a scientist; (b) he apparently experimented on patients without their permission or knowledge despite their trust, and the trust of the Infirmary, in him as a physician, thus damaging the reputation of the Infirmary; (c) according to reports it is not clear that he published all the pertinent data of his testing, in particular the negative results; and (d) testing unproved drugs could have harmed patients.

ST could have mitigated or avoided these conflicts of interest by one of several methods including: (1) disclosure of his financial interests; (2) not acting on these financial and personal interests; and/or (3) removing himself altogether from the conflict situation, such as by divesting of the financial interest or exempting himself from the research project. ST did none of these.

III. ETHICAL PROBLEMS WITH CONFLICTS OF INTEREST

From a moral point of view, there are at least three kinds of criteria for evaluating what is wrong with acting in disregard of a conflict of interest. (1) One can judge ignoring a conflict of interest in terms of professional standards, codes, or laws, finding out whether codes, standards, or laws were violated. (2) One can evaluate the conflict in terms of its foreseeable positive or negative outcomes, determining whether ignoring the conflict of interest in question created or contributed some harm, for example, by affecting scientific judgment, by biasing research or results, or by harming institutions or individuals who are affected by that conflict. (3) One can evaluate acting without regard to a conflict of interest from the point of view of the act itself, judging the activity according to whether it violated a moral rule. Let us consider these three criteria in more detail.[14]

Acting in disregard of a conflict of interest usually entails a violation of a social norm, a professional standard, or a code. Sometimes acting without regard to conflicts of interest even violates the law. Because of their membership in a profession one expects scientists, physicians, and engineers to meet the standards of their profession as spelled out in professional codes of ethics and institutional guidelines for research. When in the course of scientific research those standards are compromised for other nonprofessional interests, the person in question violates the norms of the profession. ST was accused of violating some of the norms of his profession as a scientist and as a physician. He may not have faithfully followed well-defined procedures for scientific research by possibly not reporting all results. He violated the guidelines of the hospital's Human Studies Committee as well as federal regulations for protection of human subjects, which required him to get approval for all drugs to be tested. Moreover, evidence suggests that he did not always take the well-being of his patients as his primary aim as a physician; indeed, he violated norms of informed consent and could have compromised their health.

A second way we judge the morality of an action is on the basis of its foreseeable outcomes, that is, on the basis of what harmful or beneficial results could be expected to accrue. From this point of view, whatever one's motives or intentions are, we judge the morality of human actions in terms of their foreseeable outcomes, the positive or negative utility of an action itself. The best sorts of actions are ones that maximize human interests, best satisfy desires or pleasures, or minimize harms, including benefits or harms to life, health, well-being, human dignity, autonomy, or pleasure. One measures harms and benefits in terms of the qualitative and quantitative merit, long-term and short-term results, and immediate or latent satisfaction. Reasoning in terms of foreseeable outcomes is an important criterion for evaluating conflicts of interest, since disregarding conflicts of interest often reduces the reliability of scientific results, and can result in other harms as well. This harm may be to the advancement of science, to the reputation of the scientist or institution in question, or a more specific harm to some individual or individuals.

From the perspective of evaluating actions in terms of foreseeable outcomes, ST's activities did not improve his patients' illnesses, a promising research agenda was truncated and indeed sidetracked by ST's research techniques, and the reputation of Harvard Medical School, the Infirmary, and its staff and administrators suffered. Fortunately, there was no long-term harm to any of ST's patients, but under the test conditions ST established there was a real possibility of that harm. His research techniques may have affected the testing of the ointment such that its real worth was not measured, and his reports may have failed accurately to present the results of his laboratory studies. Moreover, all of these harms could have been avoided had ST recognized and disclosed his conflict of interest, and acted in a more professional manner.

A third criterion for moral assessment appeals to more general standards that are not simply an evaluation of foreseeable outcomes nor merely reflect law, codes, or societal mores. ST's experimentation did no permanent damage to his patients, and the Infirmary is recovering its reputation. So from one perspective one might argue that although the ST incident is embarrassing, in the long run, not a great deal of harm resulted from the ST affair.

Nevertheless, ST's actions set a bad precedent, a precedent one would not want other researchers to emulate for fear of graver consequences. The fact that ST appeared not fully to inform his patients and Infirmary—raising questions about his trustworthiness with his colleagues and patients—makes his action morally objectionable whether or not this created harmful results in this particular instance. This is because some actions are judged to be wrong not merely because of their positive or negative outcomes, but because they violate standards for acceptable behavior or moral rules. These rules are standards of ordinary morality that reasonable people would agree should hold for everyone, the standards of openness, truth-telling, promise-keeping, equal rights, fairness, and not causing suffering. Reasonable people advocate and defend these standards (and there are others) because it is in their long-term interest and in the interest of all of us if these rules are followed, and it is usually not in our long-term interest if these rules are habitually violated. Moral rules set the criteria for acceptable behavior by specifying how we expect others to act without making an exception for ourselves.

But where do moral rules come from? There are a number of tempting explanations. Perhaps moral rules are God-given or part of human nature. But proving that is the case involves us in religious debates that would take us far afield. Another way to derive moral rules is through a thought experiment. Imagine that you and your family and colleagues are on a spaceship heading for a distant planet you are going to populate. You are not sure how long the journey will take nor what the conditions will be when you arrive. During the journey you and the other passengers must set up the guidelines for behavior and standards for law and justice that will regulate the new colony. Since you do not know in advance what your status will be on the new planet, you are likely to choose guidelines that will give fair opportunities for all settlers including yourself and future generations. Under those conditions, you are likely to set out guidelines that specify rights to life and basic liberties, protect contracts and promises, punish lying, stealing, cheating, and murder, and provide equal opportunity for everyone to achieve their well-being. Assuming that the spaceship travelers are reasonable people, the guidelines or rules you develop will be those agreed upon by all members of the party. One can think of moral rules, then, as those rules reasonable people, in ignorance of their own circumstances or future, would agree are the best standards for their own behavior and the behavior of others.

Though moral rules are general rules, how they are interpreted depends, in part, on the context of a particular situation. Moral standards are not exceptionless because there are times when one cannot respect all moral rules or respect them equally. For instance, when one's life is threatened one may kill in self-defense. But one can override a rule or standard (e.g., that everyone has a right to life) only when one has good reasons, reasons that other reasonable people would accept as good ones (e.g., in self-defense). These are "good reasons" just because they appeal to another standard (e.g., equal rights that include *my* right to life and freedom).[15]

If the reports are accurate, ST broke a number of moral rules. As a well-educated scientist and a physician with a good reputation he elicited the trust of the Infirmary to carry out his research in a professionally acceptable manner. He also elicited the trust of his patients, who would not expect their physician to treat them as uninformed subjects for experimental drugs. He also violated a basic trust in the scientific profession. Scientists do not expect their fellow scientists to break rules of acceptable procedure, to override standard research techniques for financial interests, or to use their name to develop and profit by a drug at the expense of research and patient interests. ST broke a number of explicit and implicit promises to his employer, to his colleagues, to his patients, and to the professions of science and medicine in which he held membership. ST also did not fully disclose his testing procedures nor reveal important failures in his test results. If he published only those test results that confirmed the soundness of his ointment he withheld important information that he had a duty to disclose; this is deception.

The ST case is a clear case of someone apparently acting in disregard of a conflict of interest. However, not all cases of acting in disregard of conflicts of interest involve clear wrongdoing. Complex cases can arise when scientists deny the existence of a conflict of interest or when conflicts are not recognized and dealt with as soon as a situation arises. (See, for example, Case 1 that follows this chapter.)

IV. CONFLICTS OF COMMITMENT

Conflicts of commitment are conflicts between at least two sets of professional obligations. These situations can indirectly lead to compromised scientific or engineering judgment. Formally stated,

> "Conflict of commitment" refers to any conflict between two sets of professional obligations that cannot both be adequately fulfilled without compromising one's judgment in fulfilling one or both of them.[16]

For example, a professor who is frequently away from the lab, giving talks at conferences, is not available to adequately provide mentoring to the students. Thus, there is a conflict between two professional obligations. Conflict-of-commitment situations do not directly create bias in scientific judgment, but can nevertheless affect the quality of judgment. If the professor is pressed for time, data may not be evaluated adequately before it is published. This violates a professional obligation.

Conflicts of commitment differ from conflicts of interest because conflicts of commitment involve the distribution of focus and effort between two sets of professional obligations, rather than a conflict between professional and financial/recognition interests. Conflicts of commitment are those conflicting commitments where competing obligations prevent honoring both commitments or honoring them both adequately. Conflicts of commitment are much harder to avoid than conflicts of interest, and acting in conflict-of-commitment situations may require a reassessment of one's obligations.

One can distinguish three kinds of conflicts of commitment: role conflicts of commitment, structural conflicts of commitment, and intellectual conflicts of commitment, although, as we shall see, in specific instances these often overlap. Like conflicts of interest, conflicts of commitment raise ethical issues when acting without regard to such conflict skews one's judgment as a scientist or engineer such that the sort of judgment one makes negatively affects other scientists, compromises research outcomes, reflects negatively on one's institution or profession, and/or harms people who depend on one's professional judgment.

Role conflicts of commitment involve clashes either between two different roles, such as one's role as a parent and as a graduate student, or between two commitments within the same role. In the latter case the conflict may be between two commitments within one institution, such as one's commitment to be a good teacher and prolific researcher. Alternatively, the clash can be between a commitment to one's primary employer, or institution, and other professional obligations independent of one's primary employer, for example, when one has commitments both to one's academic institution and outside professional activities. Many of these role conflicts entail problems of time where one simply cannot adequately perform all one's role obligations. Nevertheless, even in such inevitable cases, the shortage of time can negatively affect one's research. If one acts without regard to the conflict of commitment, there is sometimes temptation to take shortcuts, to overwork when one is tired, or to borrow from another's research.

Role conflicts occur in a variety of settings. For example, engineers and scientists who work for employers other than a university often face role conflicts of commitment between demands of their employer and demands of their profession or professional code.[17] One such example is the famous Challenger case.

Morton Thiokol, a company formed by the acquisition of Thiokol by the Morton Salt Company, was the manufacturer of the solid-fuel rocket boosters for the space shuttle program including the boosters for the ill-fated Challenger, the rocket that exploded shortly after lift off on January 28, 1986, carrying six astronauts and school teacher Christa McAuliffe.[18] The Challenger disaster has been traced to a failure of the O-rings, the seals in the connecting joint between two segments of the rocket booster, to seal one of the boosters, thereby creating the environment for the fuel explosion that resulted.

According to testimony to the Rogers Commission, the commission appointed to investigate the disaster, from the very beginning of the development of the rocket boosters, Morton Thiokol engineers working on the rocket boosters had worried about the strength and flexibility of the O-ring sealing mechanisms. As early as the sixth shuttle flight heat damage to the O-rings was evident, and it became clear in subsequent launches that the secondary O-rings that backed up the primary ones were crucially important to the shuttle, because they, too, suffered erosion. Yet the secondary O-rings were reduced in diameter in subsequent design changes, they were thought by some to be redundant, and little was done to change the design of the sealing joint. After the 17th successful flight, O-ring erosion was described by a NASA official, Larry

Mulloy, as "accepted and indeed expected—and no longer considered an anomaly."[19]

Early in January 1986 Thiokol's engineers had become particularly concerned because it was evident that the behavior of the rubber O-ring material could not be accurately predicted when atmospheric temperatures were below 30 degrees Fahrenheit. In fact, the ideal launch temperature for the O-rings was 50 degrees. On January 27, the day before the scheduled launch, a launch that had been delayed several times, this concern became heightened, because the weather at Cape Canaveral was particularly cold, and colder weather was predicted for January 28.

Accordingly, Alan McDonald, project supervisor at Thiokol, and at least 14 engineers in the solid-fuel rocket unit, including Roger Boisjoly, formally protested the launch to the vice president of the Wasatch division, Joe Kilminster, who was vice president for space booster programs at Morton Thiokol, and to NASA directly. McDonald, as manager of the engineering-design team, went so far as to refuse to sign the launch go-ahead release for Thiokol, a signature necessary in order for the launch to proceed. There were three concerns voiced about the launch. First, the engineers were worried about the potential leaks of the joints sealed by the O-rings if the latter did not function properly under the predicted low temperatures of the launch. Second, some engineers were concerned about heavy weather sear recovery of the $40 million dollar boosters after launch. And third, it was speculated that the predicted presence of ice in the booster support troughs might affect the shuttle orbiter. The engineers were asked by Thiokol managers and NASA to provide scientific data that would prove that the O-rings would definitely not seal the joints. However, the engineers were unable to substantiate their intuitions on such short notice since the flexibility of O-rings had never been tested below 47 degrees F.

The engineering team at Thiokol reported to Robert Lund, the vice president of engineering. Although under pressure from NASA to override McDonald's refusal, Lund himself originally would not sign off on the launch, agreeing with his engineers that low temperatures might affect the O-ring performance. NASA, however, was anxious to launch the Challenger and confident of its success. Realizing that, Jerry Mason, to whom Lund reported, made his now infamous remark to Lund, "take off [your] engineering hat and put on [your] management hat," whereupon Lund capitulated, agreed to the launch, and Kilminster signed off for Thiokol.[20]

This case illustrates a conflict between managerial and engineering roles. According to most engineering codes of ethics, public safety is the primary concern of the engineer. This professional standard is to take precedence over other demands even when one is working for someone else. For example, the National Society of Professional Engineers' Code of Ethics states that engineers in the fulfillment of their professional duties shall "Hold paramount the safety, health and welfare of the public in the performance of their professional duties."[21].

On the other hand, for managers, efficiency, progress, and accomplishments are primary goals. Moreover, most managers do not belong to an independent professional association with a specific code of conduct to guide their performance. In the Challenger incident, on the night before the launch the Thiokol engineers were caught between a commitment to their company, Thiokol, and their professional code. The code's rule of "safety first" could not be upheld without flouting the authority of the managers for whom the engineers worked. The engineers saw their protest against the launch stymied by the managerial mentality of their superiors, and they did not see their way clearly to blow the whistle before the launch was carried out. They did not view their code as dictating an overriding obligation. Should the engineers have blown the whistle to stop top management at Morton Thiokol or to the press before the launch? Their code implied they should have; generally accepted moral rules and the possible negative consequences of a failed launch indicated that action. At the same time, engineers in this type of situation might argue that their moral commitment to their code of safety ended when they no longer had the responsibility to sign off on the launch. One could also argue that managers, too, had obligations to place engineering safety concerns over management concerns. Faced with commitments both to a professional code of safety and to management, in this case the latter took precedent.

Conflicts of commitment can also originate from the social structure of science. For example, the reward system at a university may claim to value teaching equally with research but in fact reward research proportionally more. This system affects how a faculty member allocates time and effort despite obligations to teaching. Structural conflicts of commitment often arise in interrelationships between scientists, laboratories, and scientific discovery. An important aspect in the progress of basic science is that discoveries are shared with the community of scientists. That practice includes the custom of recognizing and giving credit to the original discoverer while at the same time allowing other scientists to repeat the experiment, expand the research, and so forth. At the same time the structure of the reward/grant system is such that it is an important professional benefit to be the first to discover something or the first to publish data about a new discovery. Conflicts develop between one's commitment to being "the first" and the professional obligation to properly recognize another colleague's or student's efforts.

Third, conflicts of commitment may be intellectual conflicts where one's passion for discovery or convictions about research findings may conflict with careful methodology or judgment. For example, the excitement of a new discovery may lead a scientist to publicize that discovery before independent verification has been made. A scientist acts with disregard for an intellectual conflict of commitment when his or her interest in a hypothesis and conviction of its validity conflict with his or her methods for substantiating that hypothesis, and thus bias the results. Such biased self-deception is a pitfall against which every scientist must constantly guard. (See Case 2, which follows this chapter).

Conflicts of interest can be avoided or dispensed with in most cases; conflicting interests and conflicts of commitment usually cannot. Sometimes, however, in specific instances, it is not easy to sort out conflicting interests from conflicts of interest and conflicts of commitment. Moreover, the mere existence of these various conflicts may have an undesirable effect.

Consider a recent complex case where both conflicts of commitment and conflicts of interest exist and could be quite important.

Between 1989 and 1992, Lt. Col. RR of the Walter Reed Army Institute of Research tested the therapeutic value of gp160, a vaccine made by the Connecticut biotech firm MicroGeneSys. The drug is intended to limit levels of HIV in the blood ("viral load"), and thus, it is hoped, retard the onset of full-blown AIDS. Measurement of viral load involves a process called the quantitative polymerase chain reaction (PCR). A Walter Reed researcher (MV) conducted a new, experimental version of PCR for RR, who, speaking at a prestigious conference in Amsterdam on 21 July 1992, compared PCR results with those from 15 recipients of gp160, and called the differences in viral load "statistically significant."

However, RR was not telling the whole story: 26 people, in all, took gp160. It was also revealed that he had used questionable criteria for his statistical analysis. When Dr. M, head of biostatistics at the Jackson Foundation, reworked the data, it showed that gp160's effect was, if anything, minimal.[22] An informal inquiry on 28 August, called by RR's superior, decided that the first analysis had been rushed, due to pressure of time. and should have been done like Dr. M's. RR accepted this conclusion, which both he and MV repeated in presentations at a gathering in Chantilly, Virginia, only days later.[23]

The explanation did not prevent two other Armed Forces AIDS researchers from lodging an official complaint that RR "overstated" his results. During the Army's investigation, undertaken by Col. D., RR said that full PCR data had not reached him until 24 July 1992, and that he had consequently reported data on only the "first 15 patients who had entered the study and who had been studied for a minimum of 18 months."[24] MV, for her part, contended that she had supplied full data by 19 May. She allowed that selection of results need not be suspect, but RR stated that he had selected nothing, and was backed up on this by another colleague who had worked with him on the first analysis. Col. D accepted RR's account, without giving clear reasons for rejecting MV's.

Suspicions of data selection persist. Dr. M resigned his post in disgust at what he sees as whitewash, and unnamed Walter Reed personnel have cast doubt on the PCR method's reliability, on the Army's impartiality, on how rushed RR's first analysis really was, and on the likelihood that reputable researchers would ever use rushed analyses. Some think the conference presentation on 21 July was part of a scheme to secure a large sum of funding. RR is on the advisory board of a group called Americans for a Sound AIDS/HIV Policy. MV claims that the organization's president, a gp160 therapy enthusiast, called her on 24 August 1992, betraying familiarity with unreleased test data. Furthermore, the organization's president once conducted an investment seminar for MicroGeneSys, thus linking RR to the interests of a company whose product he tests.

Most controversially, MicroGeneSys conducted intense lobbying of several U.S. Senators to ensure that $20 million, earmarked for Army research into gp160, was added to the 1992 Defense Appropriations bill, weeks after RR's original conference presentation. RR at that time lobbied the NIH, FDA, and the Centers for Disease Control to further the testing of AIDS vaccine in pregnant women.[25] Outraged, some say envious, researchers have accused RR of trying to make gp160 look better than it is, using political influence to circumvent peer review. A blue-ribbon panel, convened by the director of the NIH, was sufficiently critical of the appropriation to have it reversed in January 1994, the money going to more general research.[26]

RR has a commitment, a professional obligation to secure funding for his work. Is that commitment compromising his scientific judgment on the significance of the test results? RR's membership in Americans for a Sound AIDS/HIV Policy indicates that he also has a commitment to an important social cause. Is that in some way affecting his interpretation of data? Second, does RR have any connections to the interests of MicroGeneSys that could compromise his judgment regarding their product, which he is testing? The fact that he spoke to the Amsterdam Conference about results that were, at best, tenuous casts doubt about his findings. The conflict-of-interest situation is certainly exacerbated by the appropriation of government funds to test gp160 in response to MicroGeneSys lobbying after the Amsterdam presentation.

These are all conflicts of interest and conflicts of commitment. We have scant evidence that RR acted with disregard to any of these conflicts. However, the existence of these conflicts raises questions about the competency of RR as a researcher, since he used experimentally questionable criteria and a small statistical base for his original discovery. The existence of a conflict of interest between RR's financial and professional interests and the existence of a conflict of commitment between RR as a researcher and RR as a fundraiser for moneys for his research bring into question the trust people have in a professional's scientific judgment. Thus, in this case the existence of these conflicts does harm to the overall scientific enterprise.

Commitments and interests may be conflicting or create conflict situations when one's discovery is sponsored by a university or company who would like to patent that discovery. This patenting would preclude sharing information with other members of the scientific community, a tradition of the scientific profession, until the patent is secured. The question does not merely concern the ownership of discovery, but also whether the interests of science are advanced when certain discoveries are patented, whether who finances the research affects its outcome, and whether the possible financial stakes engendered by the patenting become an overriding interest. The patenting and financing of genes and gene sequencing raise such issues.

Geneticists have recently begun to patent some human genes, when they have "mapped" them (i.e., established their positions along the chromosomes) and discovered their functions. On 20 June 1991, the NIH filed an application with the U.S. Patent Office for the rights to 337 human genes, studied by Craig Venter and his research team at the NIH. This was remarkable, for the genes were unmapped, their functions unknown. Furthermore, these were not even complete gene sequences. The team had selected "random clones from a collection of cDNA which correspond to active genes" and had sequenced a part of each, using machines and robots.[27] In February 1992, Venter requested rights to 2,375 other fragments, by which time the machines were sequencing 168 genes daily.

On 10 July 1992, Venter left NIH to head up the Institute for Genome Research, a privately funded enterprise. Its goal, he says, is to "do the genome project,"[28] including gene mapping and biology, besides cDNA sequencing. The initial funding is a 10-year grant from the Healthcare Investment Corporation, a venture capital group that has funded other biotech companies, and that has also created a new company, Human Genome Sciences, Inc., to turn Venter's discoveries into products. An invention or discovery must satisfy three criteria to be considered patentable: It must be novel, nonobvious, and have some utility. Venter's application failed on all three counts in September 1992.[29] NIH originally intended to negotiate a revised claim with the Patent Office, but in February 1994 dropped its patent bid altogether as "not in the best interests of the public or science."[30] However, in January 1993 Incyte Pharmaceuticals, of Palo Alto, California, filed for patent rights on 40,000 cDNA sequences. This application is still pending.

This case raises the questions of who should fund research and of what claims the funder has on the discoveries that result. Venter was originally supported by the publicly funded National Institutes of Health. He left NIH for a privately funded institution, and sought patents on gene fragments using technology developed at NIH. In such a situation sharing scientific information and materials may conflict with the potential for financial benefit when patents are granted to private for-profit institutions. NIH guidelines for the use of research developed at NIH are still unclear. If the ultimate goal of such research is to provide better diagnosis and treatment of human genetic diseases, then the question, from the perspective of positive or negative foreseeable outcomes, must be whether patenting will actually enhance or retard the rapid development of such new technologies.

Linked to the question of funding is that of intellectual property. Is the work of Craig Venter and his researchers at NIH theirs to transfer to the Institute for Genome Research and then to patent? Many ideas for gene sequencing are theirs. Yet without NIH support they could not have gotten the project started, and without funding from Healthcare Investment Corporation they cannot continue. Concerning the patenting, one might argue that only after clearly useful products and technologies have been developed might the issuing of patents for those products and technologies not conflict with scientific progress. The basic rules for granting of patents should help ensure that patents are not issued too early in the experimental development of a field, possibly retarding its progress. The case also raises an issue of the existence of a conflict of interest. The potential for financial rewards from patents could compromise one's research focus. While making money can be a part of research, one needs to be careful that the existence of financial interests does not compromise one's judgment regarding the most appropriate research approach to use in answering important questions.

V. DEALING WITH CONFLICTS OF INTEREST AND CONFLICTS OF COMMITMENT

As all these examples illustrate, conflict-of-interest and conflict-of-commitment situations do not pose simple problems. They involve questions where it is not always clear that there is a well-defined "right thing to do." Ideally, depending on the particular circumstance, conflicts of interest should be either avoided, disclosed, or ended as soon as possible. While not all conflicts of interest can be avoided, ordinarily they can be resolved. In conflict-of-commitment cases it is usually very difficult to serve all interests equally and maximally. Yet conflict-of-commitment situations, too, can usually be mitigated. How, then, should we approach these issues and their resolution?

The disclosure of conflicts to the person or group relying upon the judgment in question has often been considered the most appropriate approach to these issues. This of course depends on the assumption that the scientist in question recognizes and acknowledges that a conflict situation exists in a particular instance. Guidelines being developed are intended to help investigators recognize a conflict-of-interest situation, and they are often designed to prevent investigations from getting into such a situation. For example, Johns Hopkins and other universities have prohibitions on researchers holding equity in companies with whom they do consulting work or from whom they receive research support. Hopkins recently reversed this policy, indicating that researchers can hold stock but ordinarily may not sell it until two years after the product on which they worked goes on the market.[31]

The Association of American Medical Colleges (AAMC) has produced its own guidelines for conflicts of interest and commitment.[32] They indicate that each institution should have its own procedure for disclosure and review, and they recommend several levels of review beginning with the department chairperson. The guidelines suggest that each institution develop its own list of possible conflict situations but they also give a list of specific situations that should be viewed as particularly problematic. These include faculty members doing research on products from companies in which they or immediate family members have financial interests, using students to perform work for a company in which the faculty member has financial interests, or unauthorized use of privileged information acquired from professional activities. The AAMC guidelines also recommend that each institution develop clear standards for faculty obligations to that institution to prevent conflicts of commitment.

Journals have also begun to ask for disclosure, and a very broad approach has recently been taken by the journal *Science*. Every manuscript there must now contain "any information about the authors' professional and financial affiliation that may be perceived to have biased the presentation."[33] Not only authors but reviewers, editors, and news writers are asked to inform the journal of

> any potential conflict of interest that might consciously or unconsciously bias their opinion in refereeing or writing a paper. That information could include

such items as financial interest, work in the reviewer's own laboratory that conflicts with or competes with the paper being reviewed, or strongly held intellectual, religious or social convictions when relevant.[34]

The information received is kept confidential, and if the editor thinks it is important, a comment is included in the author credits, and the author is consulted before publication. However, *Science*'s policies have not escaped criticism. According to one researcher,

> *Science* and other journals imply that authors' affiliations, funding sources, financial interests, intellectual passions, and perhaps even sexual orientation or religion should be somehow taken into account when one reads a paper . . . these policies are counterproductive; by shifting the attention of readers away from content, journals are encouraging ad hominem evaluations and thereby reducing the overall objectivity of scientific discourse. These policies are also ethically questionable, because they impugn authors with the implied accusation of wrongdoing without evidence and without recourse.[35]

It should be understood that disclosure does not resolve or end a conflict-of-interest situation nor exonerate those who may have acted in disregard to the conflict. "What it ends is the passive deception of allowing one's judgment (or other service) to appear more reliable than it in fact is."[36] In some cases it is also important to change the situation so as to avoid the conflict or end it. The guidelines universities are developing to recognize, monitor, and resolve conflicts are essential for accomplishing that goal.

Granting agencies have also begun to develop guidelines for identifying conflicts of interest. The Howard Hughes Medical Institute does not allow its grant holders to have any "significant" equity in companies related to the scientist's research efforts. The meaning of "significant" is not clear, and Hughes makes its decisions on a case-by-case basis.[37] NIH and NSF do not yet have regulations in place for determining financial conflicts of interest on the part of their supported investigators. Initial indications are that both agencies will allow the investigator's own intuition to determine whether a conflict of interest exists based on the information the investigator discloses. The intent of any regulations will be to discourage researchers from having financial interests in companies whose products they are evaluating.

Granting agencies have established conflict-of-interest guidelines for proposal reviewers. NSF requires applicants to list collaborators during the previous four years and graduate and postgraduate advisors and advisees. This information is then used to "identify potential conflicts or bias in the selection of reviewers."[38] The International Science Foundation instructs grant reviewers, "if you have any relationships with the institution or the persons submitting the proposal, please consider whether they could be construed as creating a conflict of interest for you."[39] If a reviewer answers yes, they are to check a box and describe their conflict of interest. Ordinarily the agency will request the review even if a conflict is thought to exist. The U.S. Department of Agriculture asks applicants to list collaborators and co-authors over the past five years to help in reviewer selection, but does not regard any other scientists in the applicant's research area as being in conflict of interest. Agency reviewers are asked to disqualify themselves if they have been collaborators, co-authors, thesis or postdoctoral advisors, graduate students, or postdoctoral associates of the applicant during the past five years. Reviewers are also to disqualify themselves if they have "an institutional or consulting affiliation with the submitting institution, applicants or collaborators or will gain some benefit from the project, financial or otherwise."[40]

Professional organizations are also producing conflict guidelines. The American Society of Mechanical Engineers has developed conflict-of-interest guidelines that must be signed by all staff, officers, and committee members. It has changed its procedures for handling interpretations of this code. These must now be reviewed by at least five people and made available to the public through their publications. Responses to these interpretations are also published and an appeal procedure for questioning code interpretations is in place.[41]

Despite these guidelines, in both journal submissions and grant review much of what constitutes a conflict of interest is left to the individual scientist. While codes, standards or guidelines are important, they alone cannot solve problems of conflict of interest or conflict of commitment facing the individual scientist or engineer. So further guidance is needed—thus our appeal is to commonsense morality. Here the guidelines for evaluation are less clear. We suggest three approaches. First, we can ask a number of questions that guide our evaluation of a specific conflict of interest or conflict of commitment:

1. Is this the kind of conflict of interest or conflict of commitment an impartial scientist or engineer would find acceptable, all things considered? In particular, will the conflict create a significant temptation for the research to rend a biased scientific judgment? Would such actions be acceptable to colleagues or to the public? In many of the cases the answer to all these questions is "no."

2. What kind of precedent would acting in disregard to this conflict of interest or conflict of commitment set? Is this a conflict peculiar to this situation or would accepting financial and personal considerations set a standard for similar cases?
3. What are the interests and expectations of parties involved or affected by this situation? Can I make my interests known to them and receive their approval?
4. The publicity test: Can this conflict of interest be made public and defended in a public forum?
5. What kind of institutional structure, accountability, procedure, or absence of constraint might have contributed to the conflict of interest or conflict of commitment? Are there structural or societal factors that must be taken into account? How might one change any of these factors to try to avoid similar dilemmas in the future?

A second set of questions need to be asked from the point of view of foreseeable outcomes. Who is harmed and who benefits if one acts in disregard to this conflict of interest or conflict of commitment? Are there trade-offs of costs and benefits that can be either avoided or, alternately, justified by long-term benefits?

Third, we can test the conflict of interest in question against moral rules: openness and avoiding deception, keeping one's promises, respecting basis rights, avoiding creating harms, and treating people fairly. Here we ask the question, does conflict of interest violate any moral rules?

Finally, if the conflict of interest or conflict of commitment in question does not pass these tests, that is, if it does not conform to codes and standards, if it does not meet the precepts of commonsense morality, and/or if acting upon it, on balance, increases harms, we are faced with the question: What should one do? What one should do, obviously, is to clear up or remove the conflict. But what specifically one should do depends on the context and facts of the situation. There are several options.

1. Disclose the conflict.
2. Divest oneself of interests that threaten independent scientific judgment.
3. Withdraw from the situation altogether, or, in some cases of evaluating research or manuscripts, do not render a judgment.
4. Appeal to laws, rules, or policies that accommodate clashing interests.

5. Change institutional procedures or policies. For example, at the time of the Challenger incident, Morton Thiokol had no mechanisms in place to expedite considerations of safety concerns, nor had it trained its managers to place engineering safety commitments ahead of managerial commitments. So the inability of Thiokol engineers to halt this launch is partly a failure of the structure of the company in which they worked, as well as the "go/no go" process of decision making.
6. Change expectations of involved parties. For example, some universities now encourage research liaisons with industry, but these are made public, and explicit guidelines are often in place so that there can be no question of hidden conflicts of interest.[42]

V. CONCLUSION

Objectivity is an ideal—an ideal by which we judge scientific investigations and technological advances. Scientific research is a process through which one works toward the ideal using various procedures that attempt to overcome biases and other hazards to discovery.[43] In that process one tries to sort out those influences, perspectives, and pressures that preclude the achievement of that ideal. Financial interests that taint one's research projects, personal interests that interfere with one's progress as a student of science, excessive enthusiasm for one's "pet" theory, as well as issues arising from conflicts of professional commitments all work against that ideal. As a result, there are personal, professional, scientific, technological, and institutional losses that, while not always measurable in the short run, lead to a net loss in scientific progress that only research under the ideal of objectivity can achieve. The issues raised by conflicts of interest and conflicts of commitment, then, are crucial ones. They must be faced openly, standards must be in place to prevent some of these occurrences, and the process of moral reasoning must be fostered to help individuals and institutions facing these issues deal with them rationally and effectively.

Notes

1. Research for the cases in this chapter was done by Jean McDowell and Liam Harte of Loyola University Chicago, to whom we are greatly indebted. This chapter benefited from comments by other members of the consortium, in particular, Stephanie Bird, Deni Elliott, Bernard Gert, Judith Swazey, and Vi-

vian Weil, as well as Rachelle Hollander and Michael Davis. Its shortcomings are unfortunately our own.

2. This view is often attributed, probably erroneously, to Francis Bacon.

3. As reported in the *Boston Globe,* October 19, 1988, p. 17.

4. P. G. Gosselin, "Flawed study helps doctors profit on drug," *Boston Globe,* October 19, 1988, pp. 1, 16; D. Kong, "Chargles against two dropped," *Boston Globe,* April 3, 1992, p. 2.

5. See W. Booth, "Conflict of interest eyed at Harvard," *Science* 242(1988):1497–99; P. G. Gosselin, "Flawed study"; P. Gosselin, "The system failed in drug research probe," *Boston Globe,* December 5, 1988, pp. 29–33; T. L. Kurt, "FDA issues concerning conflicts of interest," *IRB Report* 12(1990):1.

6. See N. Bowie and R. Duska, *Business Ethics* (Englewood Cliffs, N.J.: Prentice Hall, 1990).

7. R. S. Downie, *Roles and Values* (London: Methuen, 1971).

8. See, for example, Association of American Medical Colleges, *Guidelines for Dealing with Faculty Conflicts of Commitment and Conflicts of Interest in Research* (Washington, D.C.: Association of American Medical Colleges, 1990).

9. See, for example, Association of American Medical Colleges, *Guidelines;* Harvard University, *Principles and Policies that Govern Your Research and Other Professional Activities* (Cambridge, Mass.: Harvard University Faculty of Arts and Sciences, 1992).

10. K. J. Rothman, "Conflict of interest, the new McCarthyism in science," *Journal of the American Medical Association* 269(1993):2782.

11. Ibid., 2782–84.

12. See, for example, Association of American Medical Colleges, *Guidelines,* 6.

13. For a more rigorous definition of conflict of interest see M. Davis: A person P1 has a conflict of interest in role R if and only if:

1. P1 occupies R.
2. R requires exercise of (competent judgment) with regard to certain questions Q;
3. A person's occupying R justifies another person relying on the occupant's judgment being exercised in the other's service with regard to Q;
4. Person P2 is justified in relying on P1's judgment in R in regard to Q (in part at least) because P1 occupies R; *and*
5. P1 is (actually, latently, or potentially) subject to influences, loyalties, temptations, or other interests tending to make P1's (competent) judgment in R with regard to Q less likely to benefit P2 than P1's occupying R justifies P2 in expecting.

From M. Davis, "Conflict of interest," *Business and Professional Ethics Journal* 1(1982):17–27; see also T. Carson, "Conflicts of interest," *Journal of Business Ethics* 13(1994):387–404. Carson wrote:

> A conflict of interest exists in any situation in which an individual (*I*) has difficulty discharging the official (conventional/fiduciary) duties attaching to a position or office she holds because either: i) there is (or

> *I* believes that there is) an actual or potential conflict between her own personal interests and the interests of the party (*P*) to whom she owes those duties, or ii) *I* has a desire to promote (or thwart) the interests of (*X*) (where *X* is an entity which has interest) and there is (or *I* believes that there is) an actual or potential conflict between promoting (or thwarting) *X*'s interests and the interests of *P*.

See also N. Luebke, "Conflict of interest as a moral category," *Business & Professional Ethics Journal* 6(1987):66–81.

14. See P. Wells, H. Jones, and M. Davis, *Conflicts of Interest in Engineering* (Dubuque, Iowa: Kendall/Hunt, 1985), 5, for an outline of these three perspectives.

15. See Adam Smith, *Lectures on Jurisprudence,* ed. R. L. Meek, D. D. Raphael, and P. G. Stein (New York: Oxford University Press, 1978), especially A(i.ii., 12–14); and B. Gert, *Morality* (New York: Oxford University Press, 1988), especially chapters 1–7.

16. See, for example, Association of American Medical Colleges, *Guidelines,* 4; and *The Harvard Program in the Practice of Scientific Investigation* (Cambridge, Mass.: Harvard University Press, 1992), 34–35.

17. See, for example, the National Society for Professional Engineer's Code of Ethics for Engineers, which states that "Engineers, in the fulfillment of their professional duties shall: 1. Hold paramount the safety, health and welfare of the public in the performance of their professional duties" (NSPE publication no. 1102, revised January, 1987).

18. We use the word "was," because in 1989 Morton divested itself of the rocket booster division of Thiokol, and Thiokol is now an independent company.

19. T. E. Bell and K. Esch, "The fatal flaw in flight 51-L," *IEEE Spectrum* 24(2)(1987):43.

20. For various reports on the case, see M. M. Waldrop, "The Challenger disaster: Assessing the implications," *Science* 231(1986):662–64; R. J. Smith, "NASA had warning of potential shuttle accident," *Science* 231(1986):792; and Smith, "Thiokol had three concerns about the shuttle launch," in the same volume, p. 1064. See also *Aviation Week and Space Technology,* "Ruptured solid rocket motor caused Challenger accident," 2/10/86, p. 19; "Morton Thiokol engineers testify NASA rejected warnings on launch," 3/3/86, p. 18; "Astronauts Express Opionions on Shuttle Launch Safety," 3/17/86, p. 28; "O-Ring documentation misled managers," 3/24/86, p. 25; "NASA investigating claims that Morton Thiokol demoted engineers for disclosing launch opposition," 5/19/86, p. 29; "Rogers Commission Report Cites NASA Mangement Deficiencies," 6/9/86, p. 16; "Rogers Commission charges NASA with ineffective safety program," 6/16/86, p. 18. See also T. E. Bell and K. Esch, "The fatal flaw," 43.

21. See, for example, the National Society of Professional Engineers' Code of Ethics for Engineers, reprinted in *Ethical Issues in Professional Life,* ed. Joan Callahan (New York: Oxford University Press, 1988), 460–61; our emphasis.

22. J. Cohen, "Army investigates researcher's report of clinical trial data," *Science* 258(1992):883.

23. J. Cohen, "Army clears Redfield—but fails to resolve controversy," *Science* 261(1993):825.

24. Ibid, 824.

25. J. Cohen, "Lobbying for an AIDS trial," *Science* 258(1992):539.

26. J. Cohen, "Peer review triumphs over lobbying," *Science* 263(1994):463.

27. L. Roberts, "NIH gene patents, round two," *Science* 255(1992):912–13.

28. A. Anderson, "Controversial NIH genome researcher leaves for new $70 million institute," *Nature* 358(1992):95.

29. L. Robert, "Rumors fly over rejection of NIH claim," *Science* 257(1992):1855.

30. C. Anderson, "NIH drops bid for gene patents," *Science* 263(1994):909.

31. C. Anderson, "Hughes' tough stand on industry ties," *Science* 259(1993):884–86.

32. Association of American Medical Colleges, *Guidelines,* 1990.

33. Information for contributors, *Science* 259(1993):41.

34. D. E. Koshland, "Simplicity and complexity in conflict of interest," *Science* 261(1993):11.

35. K. J. Rothman, Letter to the editor, *Science* 261(1993):1661.

36. Wells, Jones, and Davis, *Conflicts of Interest,* 23.

37. See Anderson, "Hughes' tough stand."

38. National Science Foundation, Instructions for completing biographical sketch of principal investigator, NSF Form 1362, 1994.

39. International Science Foundation, Long-term research grants, peer review rating form, 1994.

40. U.S. Department of Agriculture, National Research Initiative competitive grants program, Guideline for reviewers, 1994.

41. C. W. Beardsley "The hydrolevel case—a retrospective," *Mechanical Engineering* 106(1984):66–73.

42. For example, ARCH Development Corporation of The University of Chicago and Argonne National Laboratory.

43. See Rothman, "Conflict of interest," 2783; and K. Popper, *The Open Society and Its Enemies* (Princeton, N.J.: Princeton University Press, 1966), especially 217.

CASES FOR CONSIDERATION

CASE 1[1]

Dr. DS, a well-known Harvard neurologist, studies the biological mechanisms of Alzheimer's disease. Dr. S is a frequent author of both professional and semipopular reviews on research in this field; within a two-year period Dr. S's reviews on Alzheimer research appeared in *Science, Neuron,* and *Scientific American.* Each of these reviews discussed the work of a new biotech company doing research aimed at developing diagnostic tests or treatment models for Alzheimer's. Dr. S did not mention in any of these reviews that he was the scientific founder and one of the largest shareholders in the biotech firm. When questioned about his ties to the firm, Dr. S stated that it did not occur to him to disclose that information and that no one asked if he had any ties to the firm mentioned. In fact, Dr. S saw nothing wrong in his actions, stating, "I felt that I was doing my usual academic job of writing a review." He pointed out that he also mentioned the work being done at several other biotech companies in which he has no financial stake.

Questions for Discussion

1. Is there any evidence that Dr. S lied about the value of the research at the new biotech company?
2. Why was he questioned about his ties to that company?
3. How could Dr. S have relieved the concerns raised? Is it necessary for him to do that?

CASE 2[2]

RM, a feisty 74-year-old archeologist, is a model of the committed scientist. For as long as anyone can recall, he has been trying to disprove the "Clovis-first" theory, which states that the earliest human settlements in the New World are 12,000 years old, at most. The date is pegged to unusual stone weapons that were found at Clovis, N. Mex., in the 1920s, and, since then, many scientists have underwritten the paradigm as a matter of course. So far, RM's efforts to undermine confidence in it have not impressed traditionalists, although he remains very well-respected in his field. RM regards the general resistance to his work as an "ingrained conservatism" that is only to be expected of those who are unwilling to admit that they have wasted "time and effort building up evidence" for their view.

Nowadays, RM thinks that he has "incontrovertible proof" that would establish a 30,000-year antiquity for human settlements in North America. "This is the one that's going to finish off the skeptics," he told the *Washington Post.* "This time we knew exactly what kind of evidence it was going to take to convince people." His proof consists of an impressive array of objects collected recently at Pendejo Cave at Fort Bliss, N. Mex. There are "500 objects" of stone—including varieties that are foreign to the cave—that RM believes were chipped by humans; a 35,000-year-old buffalo bone, showing marks of human chopping; eight hearth-remains; an animal toe-bone with a projectile point; and human fingerprints on clay dated at 30,000 to 35,000 years old, along with a human mongoloid hair of about the same age. Thirty-two neatly chronological carbon-14 dates, from distinct levels of the cave, establish a context for the objects. RM also says that "10 or 15 other experts" have authenticated his findings.

RM's eminence (he is a member of the National Academy of Sciences, and the scientific director of the Andover Institute for Archaeological Research) makes it rash to dismiss his claims out of hand, but even some who support him harbor reservations about his presentational style. There are two main worries: first, RM's habit of predicting, in the lay press, the results of his research; second, the "generosity" of his standards for pronouncing an object to be an artifact, and therefore of interest to archaeologists of human settlements. AB, of the University of Alberta in Edmonton, considers himself to be a friend and ally of RM, but is nonetheless skeptical of Pendejo's significance. RM "may be able to demonstrate his case," says AB, "but I don't think he has all the evidence to do it yet." He thinks that RM was "just premature in announcing . . . that this is going to shake the universe."

Another cautious admirer is JA of Mercyhurst College in Erie, Pennsylvania. He excavated the Meadowcroft Rock Shelter in Pennsylvania, which many consider the most credible pre-Clovis candidate site of all. "Judging by what I have seen to date in print of Pendejo," says JA, "I don't think many people—except other disciples of remote antiquity—are ready to believe that the site is as old as he says it is, whether or not he has human hair." The central question is: has RM's crusading spirit made him overestimate the strength of his evidence?

Questions for Discussion

1. Is there evidence of wrongdoing by RM?
2. What is the nature of the conflict of commitment in this case?
3. What resolution would you suggest for the perceived conflict of commitment in this case?

Notes

1. M. Barinaga, "Confusion on the cutting edge," *Science* 257(1992):616.
2. E. Marshall, "When does intellectual passion become conflict of interest?" *Science* 257(1992):620–23.

Scripps Backs Down on Controversial Sandoz Deal

CHRISTOPHER ANDERSON

The Scripps Research Institute and Sandoz Pharmaceuticals Corp. have succumbed to pressure from Congress and the National Institutes of Health (NIH) to renegotiate a controversial agreement that would have given the Swiss company first rights to nearly $1 billion in federally funded research. The proposed agreement would have given Sandoz broad control over Scripps' research and researchers and allowed the company to review the grant proposals and consulting agreements of scientists at the institute.

Richard Lerner, president of the La Jolla, California, institute, told *Science* last week that the offending language will be removed. But he said Sandoz still intends to invest a total of $300 million over 10 years, beginning in 1997, and after a 10-year agreement between Scripps and Johnson & Johnson expires.

The unprecedented agreement was attacked last week at a hearing conducted by Representative Ron Wyden (D-OR) at which NIH Director Bernadine Healy called the agreement a "dangerous" aberration. Healy said NIH has surveyed 375 university-industry agreements, and none of them contains the most controversial provisions of the proposed Scripps-Sandoz deal. Nevertheless, she said NIH will develop guidelines for all universities negotiating such agreements with industry.

The proposed contract, which Scripps announced last fall (*Science,* 4 December 1992, p. 1570), had been expected to be signed formally by

1 July. But that was before Wyden, whose investigations into the price of new drugs have led him to examine arrangements between government-funded research institutions and industry, picked up on press reports, criticizing the unprecedented scope of the deal. In March, Wyden held a hearing on the agreement at which Healy criticized it as potentially destructive and possibly illegal. At that time, Scripps defended the contract and argued that NIH's lawyers had misinterpreted it.

Now, after 3 months of additional review, NIH has stepped up its criticisms. In a six-page letter sent to Scripps on 11 June, NIH's legal advisor said the agency ". . . is concerned that the agreement unduly restricts scientific research supported by the government, appears to be contrary to the letter and spirit of the Bayh-Dole Act (a 1980 law that, for the first time, allowed industry to acquire the rights to federally funded research done by universities and private research institutions) and may impinge unreasonably upon scientific freedom . . ."

At last week's hearing, Healy told Wyden that she found it "not at all surprising that one (newspaper) columnist called the agreement 'a leveraged buyout' of a taxpayer-supported institution." She called the proposed contract an "aberration, by virtually every measure" and a "dangerous exception" to the standard arrangement with industry, adding that she was concerned that the agreement may "tempt rash, wholesale indictments" of the otherwise productive system in place to transfer federally funded technology.

Healy said that NIH was prepared to restrict all of Scripps' future NIH grants to retain government co-ownership of technology, a step that would effectively scuttle the deal with Sandoz. Scripps declined to testify, fearing what one official called a "witchhunt." Instead, the company said in a statement that its agreement is "not only completely consistent with the letter and spirit of the Bayh-Dole legislation, but also is in the best interests of science and the public."

Nevertheless, although Scripps still disagrees with much of NIH's interpretation of the contract, Lerner told *Science* that the institute plans to delete or amend "each and every one" of the elements that are controversial (see table). "We never had any intention to do the kind of things that NIH says the agreement permits," he says, but "we're happy to take out of the agreement anything that NIH finds egregious."

Although the changes are expected to be worked out over the next several months, Lerner says that one major shift will be to eliminate the language giving Sandoz first rights to all research at Scripps, leaving it with the rights only to the research it directly funds. Scripps is also expected to remove clauses giving Sandoz representation on its board of trustees and control of a joint Sandoz-Scripps scientific council. The

Table 1.
Scripps-Sandoz: Some Offending Passages

Initial Agreement	**NIH's Concerns**	**Expected Changes***
Sandoz would provide Scripps "general funding" for research of its choice in return for an exclusive worldwide license of all Scripps inventions related to medical or manufacturing products, excluding existing research agreements with third parties.	The agreement appears to restrict competition and exclude organizations other than Sandoz from having reasonable access to Scripps' technology.	About half of the Sandoz funding will go to a fund to which Scripps researchers would apply for grants. Sandoz will receive rights only to that research. The rest of the money will go to a general fund to support infrastructure and recruiting.
A "Joint Scientific Council" with a one-person Sandoz majority would monitor the relationship and have broad authority to influence the direction of research by Scripps scientists.	Anticompetitive effects because the council would establish procedures for reporting inventions, picking technologies to patent, and reviewing third-party consulting agreements.	Sandoz will not have a controlling majority of the panel.
Sandoz would be allowed to review invention disclosures stemming from federally funded research at Scripps before the disclosures are filed with the government.	Could delay the commercialization of inventions funded by the government. Permits Sandoz to delay products that might compete with its own products.	Deleted.
Sandoz would have the right to review proposed consultancy agreements by Scripps employees. Scripps researchers would be prohibited from accepting funding that involves licensing or option rights, unless Sandoz declines to fund the research itself. Sandoz may transfer Scripps research to its own facilities for further research and development.	Could inhibit collaboration and limit access of other companies to Scripps research. Allows Sandoz to review proposed agreements with other companies that may contain proprietary information.	Consulting review deleted. Restriction on third-party funding made moot by new grant mechanism, under which Sandoz only has rights to that research which it directly funds.
Scripps would seek a waiver to the requirement that products from federally funded research be manufactured in the United States.	Could violate federal law, especially since Sandoz is owned by a Swiss company that has large European manufacturing facilities.	Deleted.
Sandoz would be given two slots on the Scripps board of trustees and would be allowed to terminate the agreement if Scripps appoints a chief executive officer unacceptable to Sandoz.	Level of involvement would permit Sandoz to monitor and perhaps restrict competition in the commercialization of products form Scripps research.	Deleted.

*Details are still being negotiated.

council will be recast as a review panel, comparable to an NIH study section, to which Scripps scientists would apply for funding.

Lerner says that these changes will not reduce Sandoz's financial contribution, which will be divided between a grants program and a "no strings attached" general fund to support Scripps' infrastructure and recruiting. When the contract was first drawn up last year, he says, the two parties had not yet established a relationship of "trust" and the resulting language was conservative. "In the absence of trust," he says, "it's an issue for the lawyers."

Wyden staffers and NIH officials said that they were encouraged by Scripps' promise to modify the contract but withheld final judgment pending review of the revised agreement. Scripps gave NIH a copy of the original contract only 2 days before the first Wyden hearing and declined to make it publicly available. But last week, Scripps and Sandoz agreed to disclose it after removing some financial details. They planned to deliver a copy to the Wyden committee early this week.

The flap has already led NIH to decide to make its policies on industry research agreements more explicit. Healy announced at last week's hearing that a task force, created earlier this year to review the commercialization of intellectual property rights from NIH-supported extramural research, will draft guidelines for all such agreements.

The task force, composed of NIH scientists, program officers, technology-transfer specialists, and lawyers, has already surveyed more than 100 institutions involved in research agreements with industry. Its work, Healy said, will be published in the *Federal Register* "not as rules, but as recommendations our grantees can consider when negotiating such agreements."

Learning from the Scripps-Sandoz ordeal and NIH's own painful experience in writing conflict-of-interest guidelines, the task force will solicit public input before completing its work.

7

Institutional Responsibility

EDWARD BERGER AND BERNARD GERT

I. INTRODUCTION

In 1992 the National Academy of Science Panel on Scientific Responsibility and the Conduct of Research formulated 12 recommendations intended to "strengthen the research enterprise and clarify the responsibilities of scientists, research institutions and government agencies in this area."[1] Seven recommendations included discussion of responsibilities of the research institution and focused on three issues:

1. First, research institutions have the responsibility for fostering the integrity of the research process. This includes providing "an environment, a reward system and a training process that encourage responsible research practices." In addition it requires discussion of both the procedures for reporting misconduct and of "information about relevant laws and regulations that govern misconduct in science."
2. Second, research institutions "should adopt a single consistent definition of misconduct in science" and "have policies, protocols and procedures that ensure appropriate and prompt responses to allegations of misconduct in science should they arise."[2] These procedures must yield thorough investigation while protecting interests of the accused. The panel stressed the need for "administrators running the programs who are experienced, dedicated and prepared to devote time to misconduct issues."
3. Finally, the institution must participate in dealing with the consequences of misconduct inquiries and investigations. This may include imposing sanctions, when there is a finding of misconduct, or remedial actions, in the event of a no-misconduct finding. Remedial actions entail, on one hand, dealing with "scientific slop-

> piness, incompetence, poor laboratory management, and poor authorship practices" and, on the other hand, "providing protections for whistleblowers." Additionally, institutions must be responsible for the prosecution of those individuals shown to have made malicious and false allegations knowingly.

Here we clarify and interpret those recommendations. We explain what counts as misconduct and then deal with mechanisms by which research institutions can transmit the rules and practices of morally acceptable scientific behavior. Next, we analyze case studies to confront the issues, procedures, and concerns associated with the role of research institutions in handling allegations of scientific misconduct once they arise. Finally, we discuss the promotion of scientific integrity by the institution.

II. BACKGROUND

In 1989 and 1990, the U.S. Congress passed legislation requiring that institutions receiving funds from the Public Health Service (PHS) develop "an administrative process to review reports of scientific fraud in connection with biomedical or behavioral research,"[3–6] sponsored by the institution. This followed and extended an earlier statement made in 1982 by the Association of Medical Colleges[7] that proclaimed, "the Association believes that faculties and institutions have the primary responsibility to maintain high ethical standards in research and to investigate promptly and fairly when misconduct is alleged." In 1991, the National Science Foundation developed and published "Misconduct in Science and Engineering"[8] to govern response to misconduct that occurred during research activities supported by NSF. These early documents shared several features. They included the word "fraud," a legal term that had the criteria of both intent to deceive and the demonstration of damage or harm from the deception. "Fraud" was later replaced by "misconduct." In this way, the focus could be on the principal ethical concern associated with deception and less on demonstrating damage or harm.

A second common feature was that the responsibility for investigation, enforcement, and reporting fell on the institution, although specific guidelines explaining how these responsibilities should be met were left intentionally vague. Federal agencies, recognizing the diversity of institutional size, infrastructure, and wealth, were reluctant to provide a specific set of procedures for dealing with allegations of misconduct. Instead, the funding agencies would act as monitors. None of the agency regulations discussed in any detail the institution's responsibility for promoting scientific integrity through education, the articulation of policies governing mentoring, or the implementation of integrity fostering activities, such as seminars or workshops. The agencies also neglected to include any specific policy regarding either the protection of whistleblowers at institutions or the prosecution of individuals who make knowingly false and malicious allegations of misconduct.

By 1989, the Department of Human Health Services Office of Inspector General (OIG) found that while the majority of large, research-oriented institutions had adopted formal policies, as mandated, 78% of all institutions receiving PHS funds had not. The OIG concluded that this lack of compliance reflected concern about vagueness and the perceived temporary nature of the existing federal regulations. In 1989 and 1991, PHS published a final set of regulations that required institutions to adopt procedures and policies for dealing with misconduct.[9,10]

The basic federal requirements are, first, that misconduct procedures included two distinct steps: the inquiry—which is a preliminary review to determine whether misconduct might have occurred—and the formal investigation—which must lead to a judgment of guilt or innocence. Federal officials may participate in recommending sanctions such as debarment from receiving federal grants and exclusion from serving as peer reviewers.

III. WHAT COUNTS AS MISCONDUCT

If you are doing research with your own funds and only for your own enjoyment, you may, if you wish, make things up, or change the data points a little to make things come out the way that you would like. Just as with solitaire, you might prefer to win without cheating, but sometimes you just want to win. There is nothing wrong with conducting your private research as you might play a game of solitaire as long as you don't tell anyone that you discovered something new or proved some hypothesis. It is only in making a false claim that you have done something wrong.

Of course, the game of solitaire is not at all like the context in which science is conducted. Researchers are funded to conduct particular research by established conventions of good practice. Findings are communicated broadly throughout the scientific and, at times, the lay community.

Scientific misconduct is behavior that, if widespread, would seriously damage the institution of science. It has three main subdivisions:

1. Deception or knowing misrepresentation in proposing, performing, or reporting research, including the fabrication of evidence, falsification of data, and plagiarism. This is known in a shorthand way throughout the research community as FFP. This category also includes knowingly misrepresenting the strength of the research findings. The key moral principle here is that the researcher is intentionally leading one to a false conclusion. FFP are simply kinds of deception that are most likely to occur in research. These acts seriously damage the institution of science and thus are the first subdivision of scientific misconduct.
2. Attempts to prevent the reporting of misconduct or false claims, including harassment or retaliation against a person who has made a legitimate allegation of misconduct. This subdivision provides protection for those who report honest error as well as those reporting FFP. It does not count as misconduct to refrain from reporting your own errors, in most instances, although it would be keeping with the highest ideals of science to do so. Scientists have a moral obligation to report their own errors if those errors may result in people suffering some significant harm, such as if a researcher has mistakenly concluded that some drug is safe. However, when errors are only likely to result in futile research by others, there is no similar moral obligation to report. However, it is scientific misconduct to seek to prevent the reporting of your error by others.

   While the scientific enterprise has survived the widespread nonreporting of a scientist's honest errors, it could not survive attempts to prevent others from reporting them. Such attempts threaten the self-correcting principle of scientific research.
3. Obstruction of the research of others, including making malicious allegations of misconduct. Misconduct does not include honest error or genuine difference of opinion. This aspect includes attempts to destroy or damage the equipment or experiments of others, or to otherwise keep them from carrying out their research. It includes failure to provide materials necessary for others' research when there is a clear duty to do so. For example, journal policy may require that if the construction of a gene clone is reported, the clone be available upon request. This sub-

division also includes malicious allegations of misconduct. It is clear that these behaviors damage the institution of science.

IV. UNACCEPTABLE PRACTICES AND BEHAVIORS

How an institution responds to misconduct in this technical sense is governed by clear federal regulations. However, there are other research activities that are also morally unacceptable and for which institutions should also employ appropriate enforcement mechanisms and sanctions. These include (1) causing or risking serious harm to others without consent, (2) sexual harassment, (3) mistreatment of laboratory animals, and (4) knowing misuse of one's position for personal gain.

There are also some deceptive practices that are so widespread and tolerated within the scientific community that they don't count as misconduct. Practices such as puffery—the exaggeration of one's claims—and undeserved authorship don't damage the institution of science as do the more serious kinds of deception that count as misconduct. But they are still practices that should be avoided by scientists and sanctioned by institutions.

Sometimes, undeserved authorship or undeserved first authorship is not even regarded by the participants as deception. For example, a senior scientist may not realize that rewarding a graduate student who has not done significant work on an article first authorship counts as deception. The principal investigator (PI) may simply think of this act as rewarding the student for fruitless hard work on a project that the PI may have misconceived. However, the practice deceives future employers and potential colleagues about the work done by the junior person.

V. APPLYING THE DEFINITIONS

Analyzed below are two hypothetical vignettes that show how these definitions apply. Additional cases for discussion are included at the end of the chapter.

Vignette 1: The Grant Proposal

Dr. Block's laboratory holds its weekly in-house research meeting on Friday mornings. On Wednesday, students and postdocs find on their desks a photocopy of an NSF research grant proposal submitted by Dr. Gray, the head of a competing lab. Block has peer reviewed this proposal and now wishes to discuss the unpublished results that are described in the progress report and experiments proposed. Block's note, which is attached to the proposal, suggests that the proposal is not likely to be funded. The purpose of the lab discussion is to evaluate the data and approaches described so that Block's lab can avoid duplicating work done in Gray's lab.

Analysis: Photocopying Gray's proposal does not in itself constitute FFP in the strict sense. But Block has breached the confidentiality guaranteed Gray by the NSF peer review system. Even if Block claims that there is no intention to use Gray's results to plan future experiments, the content of the proposal will certainly influence what Block's lab does next. In this vignette, Dr. Block would be using his position as peer reviewer for personal gain.

Vignette 2: The Substituted Alloy

Dr. Hinkly's lab has just published a paper reporting the synthesis of a new ceramic-alloy that acts as a superconductor at high temperature. He immediately receives a request for a sample of the material from Dr. Li, a competitor working at a large corporate research lab. Hinkly is convinced that Li's lab will use the sample to complete studies more quickly than Hinkly can do in his smaller lab. But, he is compelled, by journal policy, to provide Li with the material requested. Hinkly sends Li a different ceramic-alloy, but identifies it as the requested material. He plans to "recognize" his error and send Li the real alloy in three months after Hinkly's group has completed their own studies. Hinkly plans to apologize to Li for his "error in sample identification."

Analysis: Here FFP does not apply as Hinkly is not fabricating or falsifying his own results. But the intentional mislabeling is deception and will result in falsifying Li's results. Knowingly sending the wrong sample is analogous to switching reagents in a colleague's lab, an act that is essentially a form of obstructing the conduct of research, and thus counts as scientific misconduct.

VI. SETTING INSTITUTIONAL POLICY

The content of the policy specifying how an institution will deal with allegations of misconduct is very important. The institution must first define misconduct and distinguish between true misconduct and other unacceptable practices, although the institution should have sanctions for both. The policy should explain the procedure for dealing with allegations. It is important that the developing policy have input from members of the research community, particularly students, post docs and junior faculty members. A policy is most likely to be understood and followed if members of the community are involved in drafting it.

The drafting of the policy necessarily educates those involved about the ethics of scientific research. Members of the community can also provide a comprehensive catalogue of the type of behaviors that they regard as misconduct and unacceptable practice. Scientists and engineers and often their students as well are reluctant to take time way from research activities to discuss abstract ethical issues. However, if the discussion takes place in a context that influences the writing, rewriting, or clarifying of the misconduct policy, then the relevance of these discussions to their own research becomes clear.

While the policy should include institutional sanctions for dealing with cases of unacceptable practice, federal regulation requires particular steps in dealing with technical misconduct.

VII. ANATOMY OF A MISCONDUCT CASE AND MORALLY RELEVANT CONSIDERATIONS

A. Allegations

First, an allegation is made by one or more complainants against one or more respondents. Allegations should be followed by preliminary and informal review to establish if the allegation has some reasonable basis in fact. The vast majority of cases are resolved at this stage.[11]

The institutional procedures should include a common and easily available entry point for receiving allegations of scientific misconduct. Allegations will range from vague or unsubstantiated statements of concern that may be made anonymously to those that are well documented.

Ethical Issues in Dealing with Allegations

The most challenging part of this early process is for the institutional official to distinguish between personal conflict and misconduct. This step can be complicated by the emotional state of the accuser and the naive belief that "it can't happen here." Anonymous allegations normally require more detail and supporting evidence than do allegations raised by identified parties because of the risk of malicious intent, but they should be investigated. Regardless of the particular situation, the critical question must be: Could there have been misconduct? Is it appropriate to look further?

A second critical issue is the dilemma faced by the person or persons making the allegation. The term "whistleblower" has a negative connotation in the scientific community.[12–14] However, reporting what appears to be a colleague's act of scientific misconduct requires courage and conviction. Whistleblowers often put themselves at risk and may lose friends, their reputations, and even their jobs.

Graduate students and postdocs are in a particular bind because of their lack of power and because of their dependence on the mentor for establishing credentials. It is ironic that institutional policies protecting human subjects and laboratory animals are usually more explicit and enforceable than policies protecting whistleblowers.

Allegations made under the condition of confidentiality raise another serious ethical concern. If an allegation turns out to be true misconduct or to proceed to the level of investigation, confidentiality cannot be promised. Institutional officials should warn those requesting confidentiality that such a promise is necessarily conditional.

B. The Inquiry

The inquiry is a more formal process of information gathering in which a determination is made concerning the availability of credible evidence that scientific misconduct might have occurred. The inquiry is "designed to enable a distinction to be made between allegations of scientific misconduct that have no merit and those that deserve further review."[15]

One or more specific allegations are chosen for examination by a panel selected by the institutional official. The inquiry must be carried out in a fair, confidential, impartial, expert, independent, and expeditious manner, and the integrity of the process cannot be contaminated by issues of bias or conflict of interest. By the time that an inquiry has begun, the respondent has been notified of the relevant details of the allegation with a clear statement of what is to follow. The inquiry panel is provided with a clear charge, time, other resources needed for the inquiry, and access to legal council and indemnification.[16]

The inquiry panel reviews evidence and interviews parties and other relevant witnesses. The inquiry ends with a decision by the panel as to whether a formal investigation is necessary. If the allegation is substantiated, the process moves to the investigation stage. If the allegation is unsubstantiated, the institutional official should seal the records. In some cases, the unsubstantiated allegation was made with malicious intent, which would result in a new allegation of misconduct—this time against the complainant.

Ethical Issues at the Inquiry Stage

At what point should an inquiry be initiated? Along with the facts, important features include the seriousness of the action, the emotional state of the parties, and the history of the parties with respect to previous misconduct.

The most contentious aspects of misconduct cases involve malicious allegations made with the intent of harming the respondent, and intentional retribution against whistleblowers who have made good-faith allegations. At this stage, the institution has the responsibility of protecting whistleblowers by removing them from the lab or otherwise shielding them from retaliation by the respondent.

Sequestering data, notebooks, specimens, or reagents that become evidence in misconduct cases raises serious ethical as well as legal concerns.[17] Institutions should have a policy of exclusive data ownership, regardless of the source of funding, and have organized and detailed mechanisms for providing copies of notebooks and other sequestered evidence to the affected laboratory while the misconduct inquiry proceeds. While sequestering data is a serious act and may be interpreted as a suspicion of guilt, it is in the respondent's best interest that the sequestering occur early, ideally at the moment that the respondent is informed of the allegation. The respondent and lab then are immediately absolved of any subsequent accusation that the data or notebooks were altered after an accusation was made.

C. The Investigation

An investigation is a formal review of the misconduct allegation with the purpose of drawing a conclusion as to whether the evidence supports a finding that scientific misconduct occurred. Investigations involve gathering and reviewing relevant evidence and testimony by an appointed panel and the preparation of a report that carefully documents evidence and conclusions.

Until the point of investigation, the relevant funding agency has not necessarily been informed of the allegation. However, institutions are required to promptly inform the agency if human subjects, animal welfare, or federal funds are at risk or if a felony is possible. This is called the "24 hour" rule. Once informed, the agency generally acts as monitor of the process, receiving reports concerning the investigation's outcome. Agencies, at their discretion, may run an independent investigation regardless of an institution's decision.

The investigation represents a formal, legal process. Allegations are examined using the rules of evidence and principles of due process and the investigating panel's charge includes a determination of intent.

Once an investigation is concluded a report is produced. The institutional official must inform the funding agency of its conclusion. The report will describe the gathering and evaluation of testimony and evidence, the conclusions drawn by the panel, a finding as to whether scientific misconduct occurred, the specific aspects of misconduct, the people responsible, and the basis for establishing intent. Some recommendation regarding penalties or sanctions may also be included in the report.

Ethical Issues Relating to Investigations

The investigating panel must be objective, expert, credible, and uphold confidentiality. It would seem irrational for an institutional official to select a panelist who has a clear or potential conflict of interest, but such errors have been made repeatedly.

One safeguard for the process can allow the respondent to change the choice of panelists. It is also sensible to have panels whose memberships do not overlap. That is, a panelist at the inquiry stage should not be involved in the investigation.

The panel is required to establish intent, which requires a distinction between scientific incompetence and deception. The absence of intent means the absence of technical misconduct.

D. Appeal

The respondent may request an appeal of the final determination. At the federal agency level, appeals have overturned guilty verdicts in several highly publicized cases.[18,19]

E. Penalties and Sanctions

Once the case has reached a final decision of misconduct or of a less serious offense, the appropriate institutional official determines penalties and sanctions. There are a number of institutional options, ranging from a letter of reprimand to continued scrutiny of the lab to termination of employment. At this point, the institution should notify journals, co-authors, collaborators, or even the public, with the intent of getting the record straight.

VIII. PROMOTING SCIENTIFIC INTEGRITY

Misconduct investigations are expensive and traumatic, so there are compelling incentives for institutions to develop and implement effective means of preventing misconduct and fostering scientific integrity. Such mechanisms should focus, first, on creating a climate that encourages and rewards responsible scientific practice and, second, on instituting guidelines and policies that discourage and penalize irresponsible, dishonest, or questionable research practice.

Academic and research institutions have long operated on the principles of faculty autonomy, academic freedom, and collegiality. So institutions have recognized the need for encouraging their scientists to accept the role of mentor and participate in creating an environment, reward system, and training process that encourages responsible research practices.

The faculty and institution are dependent upon one another for this to happen. Scientists who believe that their institution is passive, neutral, or muddled about what constitutes responsible scientific practice will be less successful mentors than those who work at an institution that proactively encourages its faculty to established reasoned standards of scientific conduct.

Notes

1. Panel on Scientific Responsibility and the Conduct of Research, *Responsible Science: Ensuring the Integrity of the Research Process.* Vol. I (Washington, D.C.: National Academy Press, 1993).

2. U.S. Congress, "Fraud in NIH grant programs," House of Representatives, Committee on Energy and Commerce, Subcommittee on Oversight and Investigations, 100th Cong., 2nd sess., 12 April 1989, Serial No. 100-189 (Washington, D.C.: U.S. Government Printing Office, 1989).

3. U.S. Congress, "Fraud in NIH Grant Programs."

4. U.S. Congress, "Scientific fraud," House of Representatives, Committee on Energy and Commerce, Subcommittee on Oversight and Investigations, 101st Cong., 1st sess., 4 and 9 May 1989, Serial No. 101-64 (Washington, D.C.: U.S. Government Printing Office, 1989).

5. U.S. Congress, "Maintaining the integrity of scientific research," House of Representatives, Committee on Science, Space, and Technology, Subcommittee on Investigations and Oversight, 101st Cong., 1st sess., 28 June 1990, No. 73 (Washington, D.C.: U.S. Government Printing Office, 1990).

6. U.S. Congress, "Scientific fraud, Part 2," House of Representatives, Committee on Energy and Commerce, Subcommittee on Oversight and Investigations, 101st Cong., 1st sess, 30 April and 14 May 1990 (Washington, D.C.: U.S. Government Printing Office, 1990).

7. Association of American Medical Colleges Ad Hoc Committee on the Maintenance of High Ethical Standards in the Conduct of Research, "The maintenance of high ethical standards in the conduct of research" (Washington, D.C.: Association of American Medical Colleges, June 1982).

8. "Misconduct in science and engineering; Final rule," *Federal Register* 56(93)(1991):22286–90.

9. "Responsibilities of awardee and applicant institutions for dealing with and reporting possible misconduct in science." *Federal Register* 54(151)(1989): 32446–51.

10. "Policies and procedures for dealing with possible scientific misconduct in extramural research," *Federal Register* 56(114)(13 June 1881). Notice Part VII, Department of HHS, PHS.

11. G. Taubes, "Misconduct: Views from the trenches," *Science* 261(1993): 1108–11.

12. D. Nelkin, "Whistle blowing and social responsibility in science," in *Research ethics* (New York: Alan R. Liss, 1983), 351–57. Reprinted from *Progress in Clinical Biological Research,* Vol. 128, 1983.

13. J. P. Swazey and S. R. Scher, eds., "Whistleblowing in biomedical research: Policies and procedures for responding to reports of misconduct," President's Commission for the Study of Ethical Problems in Medicine and Biomedical and Behavioral Research (Washington, D.C.: U.S. Government Printing Office, 1981).

14. E. J. R. Rossiter, "Reflections of a whistle blower," *Nature* 357(1992):434.

15. C. K. Gunsalus, "Institutional structure to ensure research integrity," *Academic Medicine* 68(9 Suppl.)(1993):S33–38.

16. C. K. Gunsalus, "On scientific misconduct in university research," *Knowledge: Creation, Diffusion, Utilization* 14(2)(1992):162–67.

17. Taubes, "Misconduct"; Gunsalus, "Institutional structure"; Gunsalus, "On scientific misconduct."

18. Department of Health and Human Services, Departmental Appeals Board, Research Integrity Adjudication's Panel, Decision: Midulas Popovic (DAB No. 1446), 1993; and Decision: Rameshvar K. Sharmal (DAB No. 1431), 1993.

19. C. Anderson, "The aftermath of the Gallo case," *Science* 263(1994):20–22.

CASES FOR CONSIDERATION

CASE 1[1]

Dr. Carolyn Phinney was a research psychologist at the University of Michigan in 1989, working under the supervision of Dr. Marion Perlmutter, in the Gerontology Institute. In a discussion with a colleague, Phinney expressed concern that Perlmutter had incorporated Phinney's research into an NSF grant application without crediting Phinney. The colleague, in turn, reported Phinney's concern to the Institute Director, Dr. Richard Adelman.

Adelman initiated an investigation immediately and compelled Phinney to take an active role as a whistleblower. By the end of the investigation, four separate panels had convened and in each case Perlmutter was found not guilty of plagiarism and theft of research material. It became known, however, that each panel appointed by Adelman contained at least one member who had been a participant in one or more of Perlmutter's grants.

In 1990, Phinney filed suit charging that Adelman had sought to discredit her and damage her reputation during and after the investigation. A jury eventually ruled in favor of Phinney, deciding that Adelman had violated the Michigan State Whistleblower Protection Act, and "that Perlmutter had committed fraud by making false promises regarding grants, authorship and employment of Phinney in order to obtain access to Phinney's research."

QUESTIONS FOR DISCUSSION

1. Could Phinney have refused to act as whistleblower after Adelman learned of the allegation from a third party?
2. In what ways were Adelman's actions questionable, in terms of professional ethics?
3. Who should be responsible for costs associated with litigation and damages?

CASE 2: THE NINNEMANN CASE[2]

In 1983 J. Thomas Condie, then a technician in the lab of Dr. John Ninnemann at the University of Utah, alleged that Ninnemann had intentionally misrepresented data in presentations of his work on the human immune response to burns. There followed two informal inquiries. The first was held by the Department of Surgery, where Ninnemann was a faculty member. Although the inquiry concluded that no misconduct had occurred, Condie was concerned by the conflict of interest involved in a department investigating the alleged misconduct of one of its own faculty. At this point Condie was forced to resign his position. Condie pursued the allegation further and a second ad hoc inquiry panel was formed. Although the panel agreed that Ninnemann had falsified data, no investigation was initiated. Instead, the institution reprimanded Ninnemann and requested remedial withdrawal of two manuscripts and formal correction of a third, which had already been published.

Dr. Ninnemann then moved to join the faculty at the University of California at San Diego and was allowed to transfer his NIH grant, provided that the USCD administration maintain vigilance over his work. However, Condie persisted and, in 1987, filed a freedom of information request to obtain material concerning several NIH grant proposals submitted by Ninnemann. This led Condie and the Justice Department to bring court action in 1989, under the 1863 False Claims Act. The assertion was that Ninnemann had misrepresented data in several grant applications submitted from UCSD that were later funded. This, Condie asserted, constituted legal fraud. In response to Condie's allegations, NIH requested that both UCSD and the University of Utah begin formal investigations. The University of Utah investigation panel concluded that Ninnemann had "intentionally and repeatedly misrepresented scientific data and that the earlier inquiries had been handled poorly." The UCSD panel found that while Ninnemann's data had been "presented in a misleading or improper manner" there had been no "intent to misrepresent data."

In the end, Condie, now supported by the Justice Department, concluded a settlement in which both the University of Utah and UCSD agreed to pay the federal government $1.6 million in damages. The grounds of the action were that both universities had defrauded the government by permitting Ninnemann to make false claims in research articles and in funded grant proposals and progress reports. In a separate settlement with the Department of Health and Human Service's Office of Research Integrity, Ninnemann lost eligibility to apply for or receive federal grants for a period of three years and agreed to retract five published articles and correct four others.

Questions for Discussion

1. What problems can you identify in the original inquiry at the University of Utah?
2. What might an institution do to avoid these problems?
3. Do you think that a settlement against both the Univeristy of Utah and USCD was justified?

4. Is loss of eligibility to recieve funding a fair penalty for the actions of which Ninnemann was accused?

CASE 3[3–7]: ELKASSABY AND MICHIGAN STATE

Maie ElKassaby was a PhD student in the lab of Dr. Jeffrey Williams, a microbiologist at Michigan State University. ElKassaby was working on a project supported by Williams' NIH grant, in collaboration with scientists at UpJohn Co. and physicians in the Sudan. Following a series of disagreements Williams withdrew his support as ElKassaby's PhD advisor. ElKassaby filed a grievance against Williams and removed the tissue samples and data she had accumulated on the project.

Although an inquiry found the grievance against Williams to be groundless, ElKassaby continued to refuse to return the lab data to Williams. Under the guidance of MSU officials ElKassaby was provided with three faculty advisors who helped her write up her sequestered data for publication, in spite of protests from Williams and the UpJohn collaborators. The paper was eventually published with ElKassaby as sole author.

Williams filed charges of scientific misconduct against ElKassaby and the MSU administrator and faculty involved in the preparation of the article. Both MSU and the NIH began independent investigations of Williams' allegations. The NIH panel in the end censored ElKassaby for her refusal to "permit access by collaborators and the PI to primary research materials and data," citing the refusal as a "breach of accepted scientific practice" and "therefore an act of scientific misconduct." They also recommended that a formal investigation be made against the three faculty members who helped ElKassaby write up and publish the data. The MSU internal panel concluded that ElKassaby had committed misconduct but would not investigate further if ElKassaby admitted her guilt and accepted sanctions. No action was taken on charges against the three faculty because, the panel felt, they had acted on the advice of, or under pressure from, MSU administrators.

Questions for Discussion

1. Under what circumstances would Williams be justified in separating ElKassaby from the lab project?
2. Under what circumstances could ElKassaby be justified in retaining the data and samples that, after all, she had analyzed?
3. What justification could MSU use to assemble the ad hoc committee for the purpose of writing up the work?
4. What other options were available to MSU administrators at the time?

Notes

1. C. Anderson, "Michigan gets an expensive lession," *Science* 262(1993):23.

2. P. Hilts, "Two universities pay U.S. $1.6 million in research fraud case," *New York Times,* 22 July 1994.

3. E. Marshall, "Fight over data disrupts Michigan State project," *Science* 251(1991):23–24.

4. E. Marshall, "New round in purloined data case," *Science* 251(1991):622.

5. E. Marshall, "MSU's misconduct move," *Science* 253(1991):1283.

6. E. Marshall, "MSU officials criticized for mishandling data dispute," *Science* 259(1993):592–94.

7. E. Marshall, "MSU coughs up partial mea culpa," *Science* 259(1993):1111.

Purpose and Scope of This Study

PANEL ON SCIENTIFIC RESPONSIBILITY AND THE CONDUCT OF RESEARCH, COMMITTEE ON SCIENCE, ENGINEERING, AND PUBLIC POLICY

CHARGE TO THE PANEL

The Committee on Science, Engineering, and Public Policy (COSEPUP) of the National Academy of Sciences, the National Academy of Engineering, and the Institute of Medicine sought to address these issues by convening the 22-member Panel on Scientific Responsibility and the Conduct of Research. The panel was asked to examine the following issues:

1. What is the state of current knowledge about modern research practices for a range of disciplines, including trends and practices that could affect the integrity of research?
2. What are the advantages and disadvantages of enhanced educational efforts and explicit guidelines for researchers and research institutions? Can the research community itself define and strengthen basic standards for scientists and their institutions?
3. What roles are appropriate for public and private institutions in promoting responsible research practices? What can be learned from institutional experiences with current procedures for handling allegations of misconduct in science?

The notes have been revised to provide the reader of this volume with full references.

In addition to outlining approaches to encourage the responsible conduct of scientific research, the panel was also asked to determine whether existing unwritten practices should be expressed as principles to guide the responsible conduct of research. If the panel members judged it advisable, they were encouraged to prepare model guidelines and other materials.

APPROACH, AUDIENCE, CONTENT

In responding to its charge, the panel chose a two-part approach intended to produce a report that would speak to all members of the U.S. scientific research community. First, the panel examined factors fundamental to the integrity of the research process, including scientific principles and research practices; changes within the contemporary research environment; and the roles of individuals, educational programs, and research guidelines in fostering responsible research practices. Second, the panel considered the incidence and significance of misconduct in science and also examined institutional approaches to handling allegations of misconduct, analyzing in addition the complex problems associated with responding to such allegations.

The panel chose this approach to emphasize positive steps that might be taken to assure the integrity of the research process in the current environment. Although many organizations are absorbed with responding to the problem of misconduct in science, institutional experience with recently adopted regulatory requirements is very new, and there is not yet a clear consensus about procedural approaches that may be necessary to address allegations of misconduct.[1] The panel did not attempt to resolve all of these matters in this report. At the same time, its selected approach is not intended to diminish the importance of related problems such as conflict of interest, the allocation of indirect costs, or federal support for scientific research, but rather to reflect the panel's judgment that integrity in the research process itself and issues arising from misconduct in science deserve critical examination and consideration on their own merits.

Although this report addresses concerns that affect the entire U.S. scientific community, the members of the panel were obliged to generalize from their own particular specialized expertise and experience. Unfortunately, it was not possible to develop a detailed description of the diverse styles and approaches of the various scientific disciplines, a description that might have conveyed the richness, spirit, and disciplinary differences that characterize U.S. science. The panel recognizes this limitation but believes that a general approach will guide specific interpretations and applications. This report should therefore be viewed as part of a comprehensive dialogue on and examination of integrity in the research process.

Finally, the panel emphasizes that this report speaks to *all* members of the scientific community, regardless of their institutional affiliation, whose research results become part of the scientific process. Although this report is addressed principally to an academic audience, the panel believes that the discussions, findings, and recommendations also have relevance for nonacademic research groups, including those in industry, and particularly those engaged in clinical trials and drug toxicology studies, as well as others whose members report research results at scientific meetings and publish in journals. Officials at research institutions also are responsible for taking educational, preventive, and remedial approaches to dealing with scientific conduct issues. All who participate in the research enterprise share responsibility for the integrity of the research process.

METHODS, DEFINITIONS, AND BASIC ASSUMPTIONS

Evaluating Available Data

The panel sought to develop a report that would address conflicting perspectives and priorities basic to enhancing integrity in the research process. An examination of empirical studies on research behaviors yielded few significant insights.[2]

The panel also concluded that existing social studies of the U.S. scientific research enterprise are not adequate to support conclusions about the relative effectiveness of various alternatives for fostering the integrity of the research process. For example, the value of formal and informal educational approaches in fostering responsible research practices has, to the panel's knowledge, not been systematically addressed. And although some research institutions in recent years have adopted formal guidelines designed to foster responsible practices, the experience with research guidelines is limited.[3]

The panel also found barriers to obtaining data on specific incidents of misconduct. Confidential institutional reports are not available if misconduct cases are under appeal or are subject to litigation, if the institutions have negotiated private settlements with the subjects of misconduct complaints, if there are findings of no misconduct, or if the misconduct has been judged to be not significant enough to warrant penalties. Those involved in handling or evaluating misconduct cases are usually not at liberty to discuss their findings. Those who have been parties at interest in misconduct cases may have a biased view of specific actions. An increasing amount of litigation in misconduct cases has further complicated the collection and analysis of primary data.

Thus many of the panel's findings and recommendations are derived from informed judgments based on discussions with persons knowledgeable about the research process and about factors that affect the contemporary research environment. The panel also met with individuals who have both knowledge of and a broad range of perspectives on the significance of the reported cases of misconduct in science. The panel's overall outlook and opinions are based on general ethical principles that are well accepted by scientists and by society.

Defining Terms—Articulating a Framework for Fostering Responsible Research Conduct

The panel defined the term "integrity of the research process" as the adherence by scientists and their institutions to honest and verifiable methods in proposing, performing, evaluating, and reporting research activities. This term is sometimes thought to be synonymous with "integrity of science," but the terms of reference are different.[4] Science is not only a body of information, composed of current knowledge, theories, and observations, but also the process by which this body of knowledge is developed. Furthermore, the scientific process is a social enterprise that involves individuals and institutions engaged in developing, certifying, and communicating research results. Throughout this report the panel focuses on the integrity of the research process as defined above.

Misconduct in science is commonly referred to as fraud.[5] But most legal interpretations of the term "fraud" require evidence not only of intentional deception but also of injury or damage to victims. Proof of fraud in common law requires documentation of damage incurred by victims who relied on fabricated or falsified research results. Because this evidentiary standard seemed poorly suited to the methods of scientific research, "misconduct in science" has become the common term of reference in both institutional and regulatory policy definitions.

However, "misconduct in science" as commonly used is an amorphous term, often covering a spectrum of both significant and trivial forms of misbehavior by scientists. The absence of a clear, explicit definition that focuses on actions highly detrimental to the integrity of the research process has impeded the development of effective institutional oversight and government policies and procedures designed to respond to such actions. Varying definitions of misconduct in science have also impeded comparison of the results of survey studies. If, for example, survey respondents apply the term "misconduct in science" to a broad range of behaviors that extend beyond legal or institutional definitions, their responses weaken the significance of reported survey results.

In order to provide policy guidance for scientists, research institutions, and government research agencies concerned about ensuring the integrity of the research process as well as addressing misconduct in science, the panel developed a framework that delineates three categories of behaviors in the research environment that require attention. These categories are (1) misconduct in science, (2) questionable research practices, and (3) other misconduct.

The panel seeks to accomplish several goals by proposing these three categories. Foremost is a precise definition of misconduct in science aimed at identifying behaviors that scientists agree seriously damage the integrity of the research process. For example, although using inadequate training methods or refusing to share research data or reagents are not desirable, such actions generally are regarded as behaviors that are not comparable to the fabrication of research data. In the same manner, sexual harassment and financial mismanagement are illegal behaviors regardless of whether scientists are involved, but these actions are different from misconduct in science because they do not compromise, in a direct manner, the integrity of the research process.

Unethical actions of all types are intolerable, and appropriate actions by the research community to address such problems are essential. But the panel believes that there are risks inherent in developing institutional policies, procedures, and programs that treat all of these behaviors without distinction. Inappropriate actions by government and institutional officials can create an atmosphere that disturbs effective methods of self-regulation and harms pioneering research activities. In particular, many scientists are concerned that the term "misconduct in science," which has been construed as including "serious deviations from accepted practices" (as currently defined in government regulations), could be defined in such a way that it could be applied inappropriately to the activities of honest scientists engaged in creative research efforts.

The panel recognizes that this framework may not satisfy all scientists, lawyers, or policymakers. Its primary purpose is to advance the quality of policy and educational discussions about distinctions between different kinds of troubling behavior within the research environment, and to allow scientists, institutional officers, and public officials to focus their attention and their efforts toward prevention on substantive issues rather than discrepancies in terminology. Thus the framework of definitions proposed in this report should be viewed as a tool for use in a sustained effort by the research community to strengthen the integrity of the research process, to promote responsible research conduct, and to clarify appropriate methods to address instances of misconduct in science. The three categories will need to be refined through continued dialogue, criticism, and experience.

In developing its framework of definitions, the panel adopted an approach that evaluates how seriously the various behaviors compromise the integrity of the research process. The panel also considered other criteria, such as intent to deceive. The panel concluded that while intention is important, especially in the adjudication of allegations of misconduct in science, intention is often hard to establish and does not provide, by itself, an adequate basis for separating actions that seriously damage the integrity of the research process from questionable research practices or other misconduct.[6]

Misconduct in Science

Misconduct in science is defined as fabrication, falsification, or plagiarism, in proposing, performing, or reporting research. Misconduct in science does not include errors of judgment; errors in the recording, selection, or analysis of data; differences in opinions involving the interpretation of data; or misconduct unrelated to the research process.

Fabrication is making up data or results, falsification is changing data or results, and plagiarism is using the ideas or words of another person without giving appropriate credit.

By proposing this precise definition of misconduct in science, the panel is in unanimous agreement that the core of the definition of misconduct in science should consist of fabrication, falsification, and plagiarism. The panel unanimously rejects ambiguous language such as the category "other serious deviations from accepted research practices" currently included in regulatory definitions adopted by the Public Health Service and the National Science Foundation.[7] Although government officials have often relied on scientific panels to define "other serious deviations," the vagueness of this category has led to confusion about which actions constitute misconduct in science. In particular, the panel wishes to discourage the possibility that a misconduct complaint could be lodged against scientists based solely on their use of novel or unorthodox research methods. The use of ambiguous terms in regulatory definitions invites exactly such an over-expansive interpretation.

In rejecting the "other serious deviations" category, the panel considered whether a different measure of flexibility should be included in its proposed definition of misconduct in science, so as to allow the imposition of sanctions for conduct similar in character to fabrication, falsification, and plagiarism. Some panel members believe that the definition should also encompass other actions that directly damage the integrity of the research process and that are undertaken with the intent to deceive. For example, misuse of the peer-review system to penalize competitors, deceptive selection of data or statistical analysis, or encouragement of trainees to practice misconduct in science might not always constitute a form of fabrication, falsification, or plagiarism. Yet such actions could, in some circumstances, damage the integrity of the research process sufficiently to constitute misconduct in science.

All members of the panel support the basic definition of misconduct in science proposed above, but the panel did not reach final consensus on whether additional flexibility was needed to address as misconduct in science other practices of an egregious character similar to fabrication, falsification, and plagiarism. These issues deserve further consideration by the scientific research community to determine whether the panel's definition of misconduct in science is flexible enough to include all or most actions that directly damage the integrity of the research process and that were undertaken with the intent to deceive.

Questionable Research Practices

Questionable research practices are actions that violate traditional values of the research enterprise and that may be detrimental to the research process. However, there is at present neither broad agreement as to the seriousness of these actions nor any consensus on standards for behavior in such matters. Questionable research practices do not directly damage the integrity of the research process and thus do not meet the panel's criteria for inclusion in the definition of misconduct in science. However, they deserve attention because they can erode confidence in the integrity of the research process, violate traditions associated with science, affect scientific conclusions, waste time and resources, and weaken the education of new scientists.

Questionable research practices include activities such as the following:

- Failing to retain significant research data for a reasonable period;
- Maintaining inadequate research records, especially for results that are published or are relied on by others;
- Conferring or requesting authorship on the basis of a specialized service or contribution that is not significantly related to the research reported in the paper;[8]
- Refusing to give peers reasonable access to unique research materials or data that support published papers;
- Using inappropriate statistical or other methods of measurement to enhance the significance of research findings;[9]
- Inadequately supervising research subordinates or exploiting them; and
- Misrepresenting speculations as fact or releasing preliminary research results, especially in the public media, without providing sufficient data to allow peers to judge the validity of the results or to reproduce the experiments.

The panel wishes to make a clear demarcation between misconduct in science and questionable research practices—the two categories are not equivalent, and they require different types of responses by the research community and research institutions. However, the relationship between these two categories is not well understood. It may be difficult to tell, initially, whether alleged misconduct constitutes misconduct in science or a questionable research practice. In some cases, for example, scientists accused of plagiarism have testified about an absence of appropriate training methods for properly citing the work of others. The selective use of research data is another area where the boundary between fabrication and creative insight may not be obvious.

The panel emphasizes that scientists, individually and collectively, need to take questionable research practices seriously because when tolerated, such practices can encourage an environment that fosters misconduct in science. But questionable practices are not equivalent to misconduct in science, and they are not appropriate subjects for investigations directed to misconduct.

Other Misconduct

Certain forms of unacceptable behavior are clearly not unique to the conduct of science, although they may occur in a laboratory or research environment. Such behaviors, which are subject to generally applicable legal and social penalties, include actions such as sexual and other forms of harassment of individuals; misuse of funds; gross negligence by persons in their professional activities; vandalism, including tampering with research experiments or instrumentation;[10] and violations of government research regulations, such as those dealing with radioactive materials, recombinant DNA research, and the use of human or animal subjects. Industry-university relationships, and the resultant possibility of conflicts of interest, also raise issues that require special attention.

In these cases, recognized legal and institutional procedures should be in place to address complaints and to discourage behavior involving forms of misconduct that are not unique to the research process. Allegations of harassment, for example, should be handled by officials designated to implement personnel or equal opportunity regulations. Allegations of misuse of research funds should be addressed by those responsible for the financial integrity of the research institutions involved. The panel concluded that such behaviors require serious attention but lie outside the scope of the charge for this study.

On some occasions, however, certain forms of "other misconduct" are directly associated with misconduct in science. Among these are cover-ups of misconduct in science, reprisals against whistle-blowers, malicious allegations of misconduct in science, and violations of due process protections in handling complaints of misconduct in science. These forms of other misconduct may require action and special administrative procedures.

Understanding Causes and Evaluating Cures

The causes of misconduct in science are undoubtedly diverse and complex. Individual scientists, institutional officials, and scholars in the social studies of science over the past decade have suggested that various factors lead to or encourage misconduct in science, but the influence of any individual factor or combination of suggested factors has not been examined systematically. Two alternate, possibly complementary, hypotheses have been advanced for considering the causes of misconduct in science and formulating methods for prevention and treatment. Many observers have explained the problem of misconduct in science as one that results primarily from character or personality flaws, from environmental stimuli in the research system, or from some interaction of both:[11]

1. *Misconduct in science is the result of individual pathology.* Misconduct in science is commonly viewed as the action of a psychologically disturbed individual. An analysis by Bechtel and Pearson (1985) of 12 cases of deviant behavior reported in the 1970s and early 1980s supported the hypothesis that scientists who engage in deviant behavior are commonly individuals who operate alone and who conceal their misconduct.[12]
2. *Factors in the modern research environment contribute to misconduct in science.* But although the "bad person" approach to explaining deviant behavior in science has had strong support within the scientific community, Bechtel and Pearson and others have questioned whether this hypothesis alone adequately explains the phenomenon of misconduct in science.

   A broad range of factors in the research environment have been suggested as possible causes of misconduct in science. Such factors include (a) funding and career pressures of the contemporary research environment (such as the pressure to publish);[13] (b) inadequate institutional oversight; (c) inappropriate forms of collaborative arrangements between academic scientists and commercial firms; (d) inadequate training in the methods and traditions of science;[14] (e) the increasing scale and complexity of the research environment, leading to the erosion of peer review, mentorship, and educational processes in science; and (f) the possibility that misconduct in science is an expression of a broader social pattern of deviation from traditional norms. In addition, it has been noted that some areas of research, such as biological and clinical research, do not yet proceed from explicit scientific laws and also make extensive use of empirical observations not related to theory. Moreover, the characteristics of certain research materials in these fields inhibit the replication of research findings as a vehicle for self-correction.

The panel has reviewed various suggestions about possible causes of misconduct in science but makes no judgment about the significance of any one factor. The panel believes that speculations about individual pathology or about environmental factors as the primary causes have not been verified; misconduct in science is probably the result of a complicated interaction of psychological and environmental factors. Moreover, although one or more such factors may contribute to specific cases of misconduct in science, the panel has not discerned a broad trend that would highlight any single factor as a clear generic cause.

Regardless of the causes of deviant behavior, the panel is concerned that some "cures" for misconduct in science would damage the research process itself. The uncertainty of evidence about external factors as causes means that recommending policy solutions for treating and preventing the problem of misconduct in science is problematic. As a result, efforts to foster integrity in the research process and to reduce the occurrence of misconduct in science should be evaluated systematically to identify steps that prove to be effective.

Notes

1. See, for example, the reports resulting from three workshops sponsored by the National Conference of Lawyers and Scientists, American Association for the Advancement of Science, and the American Bar Association (AAAS-ABA, 1989).

2. Some good examples of studies of scientific practice and the social organization of science include S. Traweek, *Beamtimes and Lifetimes: The World of High Energy Physicists* (Cambridge: Harvard University Press, 1988); D. L. Hull, *Science as Process: An Evolutionary Account of the Social and Conceptual Development of Science* (Chicago: University of Chicago Press, 1988); B. Latour, *Science in Action: How to Follow Scientists and Engineers Through Society* (Cambridge: Harvard University Press, 1987); B. Latour and S. Woolgar, *Laboratory Life: The Social Construction of Scientific Facts* (Beverly Hills: Sage Library of Social Research, 1979); E. J. Hackett and D. E. Chubin, "Scientific malpractice and the politics of knowledge," annual meeting of the Society for Social Studies of Science, Minneapolis, October 1990; and E. J. Hackett, "Science as a vocation in the 1990's: The changing organizational culture of academic science," *Journal of Higher Education* 61(3)(1990):241–79.

3. It is the panel's hope that the base of knowledge will be augmented by additional data derived from systematic evaluation of experiences in fostering responsible research practices. See also the background paper on this topic prepared by Nichols Steneck for the Panel on Scientific Responsibility and the Conduct of Research Committee on Science, Engineering, and Public Policy, in volume 11 of *Responsible Science: Ensuring the Integrity of the Research Process* (Washington, D.C.: National Academy Press, 1992).

4. A discussion of the dimensions of integrity in science is included in chapters 1 and 12 of G. Holton, *Thematic Origins of Scientific Thought: Kepler to Einstein,* revised edition (Cambridge: Harvard University Press, 1988).

5. Discussions focused initially on "scientific fraud" but encountered difficulties with the legal definition of the word "fraud." Government regulations and institutional policies have adopted terms such as "research misconduct," "scientific misconduct," and "misconduct in science," but these terms are sub-

ject to a variety of interpretations. For early discussions about the relationship between fraud and misconduct in science, see R. Andersen, "The federal government's role in regulating misconduct in scientific and technological research," *Journal of Law and Technology* 3(1988):121–48. See also the discussion of "fraud" and "misconduct" on p. 32447 in U.S. Department of Health and Human Services, "Responsibility of PHS awardee and applicant institutions for dealing with and reporting possible misconduct in science: Final rule," *Federal Register* 54 (8 Aug. 1989):32446–51. Some scientists object to the terms "scientific fraud" or "misconduct in science" because the fabrication and falsification of research results are deceptive acts that are not in themselves science. However, the social, political, and legal framework in which scientists must operate requires that we admit to the possibility of deliberate falsehoods that may masquerade as science.

6. Some institutional policies make *intention* or *deception* an explicit part of their definition of misconduct in science, whereas other policies assume, implicitly, that intention is part of the common understanding of actions, such as falsification, fabrication, and plagiarism, that constitute misconduct in science. See, for example, the definitions in the policies for addressing allegations of misconduct in science included in Volume II of this report.

Another approach considered by the panel in defining behaviors that violate the integrity of the research process was to deal only with misconduct in science and questionable research practices and to omit "other misconduct" as a category for a frame work of definitions. Although the panel chose to focus on behaviors that directly compromise the integrity of the research process, it also wanted to recognize the public dimensions of discussions about misconduct in science. Thus the panel concluded that issues such as conflict of interest, mismanagement of funds, and the harassment of colleagues on the basis of race or gender must necessarily be recognized in a framework of definitions intended to categorize behavior that adversely affects the conduct of scientific research. These forms of "other misconduct" deserve serious and sustained analysis on their own merits, but such an examination was beyond the resources and scope of this particular study.

7. U.S. Department of Health and Human Services, "Responsibilities of PHS awardee and applicant institutions for dealing with and reporting possible misconduct in science: Final rule"; and National Science Foundation, "Misconduct in science and engineering: Final rule," *Federal Register* 56 (May 14, 1991):22286–90.

8. It is possible that some extreme cases of noncontributing authorship may be regarded as misconduct in science because they constitute a form of falsification. These would include only cases in which an individual who has made no identifiable contribution to a research paper is named, or seeks to be named, as a co-author.

9. See J. C. Bailar, "Science statistics and deception," *Annals of Internal Medicine* 104 (Feb. 1986):259–60.

10. The fourth report of the NSF inspector general (NSF, 1991a) describes a misconduct case involving tampering with other researchers' experiments. This type of case would not constitute misconduct in science under the panel's definition. An allegation of this type of incident should be addressed under regulations governing vandalism or destruction of property.

11. As noted in H. K. Bechtel and W. Pearson, "Deviant scientists and scientific deviance," *Deviant Behavior* 6(1985):237–52, several leading figures in the scientific community have advocated the "disturbed individual" theory. For discussions of the impact of reward systems and social controls on deviant behavior in science, see the analysis by H. Zuckerman, "Deviant behavior and social control in science," in *Deviance and Social Change,* ed. E. Sagaran (Beverly Hills, Calif.: Sage, 1977), 87–138. For a historical perspective, see Gaston, J., "Disputes and deviant views about the ethos of science" in *The Reward System in British and American Science* (New York: John Wiley & Sons, 1978).

12. The authors concluded that the deviant behavior in these cases, usually faking scientific experiments and data, was displayed by single individuals who acted alone. They observed that many of these individuals held positions of high social status and respectability within their professions and that the scientists involved also made elaborate efforts to conceal their illegitimate behavior.

13. See National Science Board, "Report of the NSB Committee on Openness of Scientific Communication" (Washington, D.C.: National Science Foundation, 1988).

14. It has been suggested that research physicians whose sole degree is an M.D. have not been adequately exposed to the scientific methods and skills that are the foundation of a Ph.D. program.

The Definition of Misconduct in Science: A View from NSF

DONALD E. BUZZELLI

In April 1992, a panel convened by the National Academy of Sciences issued its report on misconduct in science.[1] At the end of its analysis, the report made 12 recommendations. Perhaps the most controversial of these recommendations concern the definition of misconduct that research institutions and government agencies should use when they deal with misconduct allegations. Since April, a number of science-related federal bodies and professional societies have been asked to endorse the report's recommendations. Funding agencies like the National Science Foundation (NSF) have been asked to adopt and implement them, especially the recommendations about the definition.

Some of these agencies, including NSF, already had considerable experience in handling misconduct cases. Because the Academy panel did not consult NSF about the definition, officials who work every day on misconduct cases as I do did not have the opportunity to contribute their thoughts and experience. Existing definitions can always be improved, and the report contains some useful ideas in this direction. Nevertheless, the changes the panel proposed in the definition are not

Reprinted with permission from *Science* 259(1993):584–85, 647–48, and by permission of the author.

The author is a senior scientist in the Office of Inspector General, National Science Foundation, 1800 G Street, NW, Washington DC 20550. The views expressed are the author's and are not necessarily positions of the Office of Inspector General, the National Science Board, or the National Science Foundation.

helpful and in fact would hamper the ability of research institutions and federal agencies to deal with important cases of misconduct in science. Here, I will discuss the panel's recommendations. I will also try to explain NSF's definition and show how it was applied to an important and controversial case.

THE PANEL'S RECOMMENDATION

The panel unanimously recommended the removal of "ambiguous language such as the category 'other serious deviations from accepted research practices' currently included in regulatory definitions adopted by the Public Health Service and the National Science Foundation."[2] With less unanimity, it adopted a definition limited to "fabrication, falsification, or plagiarism, in proposing, performing, or reporting research" (p. 27). In other words, it proposed that nothing should replace the "other serious deviation" phase.[3] Finally, it recommended that research institutions and government agencies adopt a single, consistent definition of misconduct in science "based on" fabrication, falsification, and plagiarism (p. 147).

The panel's reason for proposing removal of the "other serious deviation" phase was that (p. 27)

> the vagueness of this category has led to confusion about which actions constitute misconduct in science. In particular, the panel wishes to discourage the possibility that a misconduct complaint could be lodged against scientists based solely on their use of novel or unorthodox research methods.

Below, I will give some reasons for retaining language like "other serious deviation" in the definition of misconduct in science. The report gives no examples to show that confusion has occurred under the NSF definition in an actual case or that any scientists have been accused of misconduct under agency regulations because they were creative or unorthodox. The suggestion that NSF would bring such a case shows no understanding of NSF or of its misconduct procedures. The report does mention two cases in which NSF allegedly misused the "other serious deviation" phrase in a different way, by treating other kinds of misconduct as misconduct in science. These cases seem to be used as additional reasons for removing the "other serious deviation" phrase from the definition. I will take them up after discussing NSF's misconduct cases in general.

NSF'S MISCONDUCT CASES

NSF handles allegations of misconduct in science under regulations published in July 1987 and revised in May 1991. The Office of Inspector General (OIG) receives all cases that come to NSF's attention. Program officers are not informed of misconduct allegations that have been made against applicants for funding, and they may not take such allegations into consideration when processing proposals. If a case seems to require a full, formal investigation, OIG will usually allow the institution that employs the accused party to do it. The institution may make a finding of misconduct and impose its own sanction. OIG may accept the institution's investigation in lieu of an OIG investigation, may supplement the institution's investigation, or may even conduct a full investigation itself.

If OIG decides that a case seems to warrant a finding of misconduct and a sanction by NSF, it makes that recommendation to the deputy director of NSF. The deputy director arranges a hearing, if appropriate, and makes the adjudication. OIG investigates cases but cannot make findings of misconduct on behalf of NSF or impose sanctions. This is one of the protections that NSF regulations offer against abusive cases, such as the punishment of creative and unorthodox research as misconduct.

OIG was established in early 1989, so that it has over 3 years of experience in handling misconduct cases. Some numbers can be given to illustrate the case load, the types of cases received, and their resolution, but these numbers have little statistical value.[4] From the formation of OIG in early 1989 to the end of June 1992, OIG added 124[5] cases to its misconduct files. The number of cases rose rapidly at first but seems to be settling down to about 50 per year. Of the 124, approximately 70 have to do with intellectual property: plagiarism, theft of research ideas, or failure to give credit. One reason why there are so many cases of this type is that NSF program officers sometimes receive misconduct complaints from proposal reviewers and pass them on to OIG. Reviewers are especially likely to notice intellectual property problems. Although most of the misconduct cases that have received media attention involve the fabrication or misrepresentation of data, only about ten of NSF's cases are of this kind.

As of the end of July 1992, 67 of these 124 cases had been closed. Most of these were resolved without a formal investigation and did not lead to a formal finding of misconduct or a sanction. In some cases, preliminary inquiry showed that the allegation was not really about misconduct in science or that the offense that occurred was trivial. For example, a case involving a very small amount of plagiarism in a proposal may be resolved by OIG's clarifying what happened and having the applicant send a corrected proposal to NSF. In other cases, there was too little evidence to justify a full investigation or it was found that no NSF proposal or award was involved so that NSF had no jurisdiction in the matter.

Among the 67 closed cases, there were 8 in which NSF had jurisdiction and in which the university involved performed a formal investigation. Some of these cases began at the university, which then notified NSF. Others were sent to the university by NSF or another agency. Three of these investigations led to findings of misconduct by the institution, and all three were serious plagiarism cases. In a fourth case, involving data falsification, OIG disagreed with the university's finding of no misconduct but did not pursue the matter because the subject of the investigation was a foreign citizen who had permanently left the country and was not likely to apply for federal grant funds in the future. Two major cases were investigated by OIG itself without a university investigation. One case came under the "other serious deviation" provision of the definition and is discussed below; the other involved possible noncompliance with guidelines for recombinant DNA research.

OIG has sent four cases to the Office of the Director of NSF with the recommendation that NSF make its own finding of misconduct and impose its own sanction. One was the "other serious deviation" case and the other three were plagiarism cases that had been investigated at universities. In all four cases, OIG recommended that the individual be debarred from receiving federal or NSF funds for a period of time. The director's office accepted the OIG recommendations, and the cases were resolved by debarment or voluntary agreements equivalent to debarment.[6]

THE MAJOR "OTHER SERIOUS DEVIATION" CASE

Most of NSF's major cases have involved plagiarism, but one involved an "other serious deviation from accepted practices." This case deserves full discussion here because it illustrates the need for such a phrase in the definition of misconduct in science and also because the Academy report discusses it in a misleading way.

In late 1989, OIG began receiving complaints from women who had served as graduate teaching assistants in a field research project that had NSF funding. The complaints were against a senior researcher who had been taking teams of undergraduate students to a remote site in southern Mexico as part of a project in which they would observe and report on the behavior of a colony of primates. The project was supported by grants from NSF's Research Experiences for Undergraduates program. This program is intended to give active research experiences to undergraduate students, so that talented students can be attracted to research careers. The program especially tries to increase the participation of women in research.

In carrying out this project, the senior researcher was accused of a range of coercive sexual offenses against various female undergraduate students and teaching assistants, up to and including rape. These offenses occurred at and near the research site, in a private vehicle on the way to the site, and in the researcher's office, home, and car in the United States. He rationed out access to the research data and the computer on which they were stored and analyzed, as well as his own assistance, so that they were more available to those students who accepted his advances. He was also accused of threatening to blackball some of the graduate students in the professional community and to damage their careers if they reported his activities. For various reasons, this case has not been prosecuted under criminal or civil rights statutes.

OIG investigated the case itself instead of sending it to the subject's university because the university was not the grantee institution and had no involvement in the NSF grants under which this project was done. The grantee institution was a very small nonprofit institution that did not have professional staff members who were sufficiently distant from these grants and who had the expertise to conduct a difficult investigation that still had criminal implications. In any case, the researcher was no longer employed there.

The OIG investigation accumulated convincing evidence that the accusations were correct and that the subject showed a pattern of such behavior. OIG sent an investigation report to the Office of the Director of NSF with the recommendation that the subject be debarred from receiving any federal grant funds for 3 years. The case was also presented to other senior NSF management and to key members of the National Science Board, and all concurred in OIG's evaluation of it. The director's office raised the recommended term of debarment to 5 years and proposed that term to the subject. After negotiation, the subject waived his right to a hearing and agreed to a 5-year exclusion from federal support.

This was a genuine instance of misconduct in science. This case illustrates a "serious deviation from accepted practices" that is not falsification, fabrication, or plagiarism. The subject was never accused of any of those offenses. In fact, he was not accused of any kind of deception to a significant degree. For some, the only type of activity over and above falsification, fabrication, and plagiarism that might be considered misconduct in science would be some other kind of misrepresentation or deception.[7] I will discuss the view the Academy panel took of the case and then I will try to explain why NSF acted as it did.

The Academy report maintains that this case was not misconduct in science but was rather what it calls "other misconduct" (pp. 82–83 and 86). The panel did not address the specifics of this case and perhaps did not know them.[8] Instead, it treated the case as a generic instance of "sexual harassment" and argued in general that such cases are not misconduct in science. According to the report, sexual harassment is not misconduct in science because it is "not unique to the conduct of science, although (it) may occur in a laboratory or research environment." Rather, it is "subject to generally applicable legal and social penalties" and "should be handled by officials designated to implement personnel or equal opportunity regulations" (p. 29). Furthermore, the report continued, sexual harassment, sexual assault, and professional intimidation are not misconduct in science because they "do not require expert knowledge to resolve complaints" (p. 86).[9]

These general arguments have little validity. This case was not essentially a sexual harassment case, but sexual offenses were obviously at the heart of it. Such offenses can be misconduct in science, even though they are "not unique to the conduct of science." Plagiarism is also not unique to science, but it is universally regarded as misconduct in science when it occurs in a scientific setting. Similarly, the fact that there are other laws and regulations against sexual offenses would not always keep them from being misconduct in science: Even if laws were passed against fabricating data, the fabrication of data would still be misconduct in science. With regard to "expert knowledge," the amount required varies considerably from case to case. Thus, the need for expert knowledge in the resolution of misconduct cases is not a useful criterion for what is or is not misconduct in science.

Furthermore, the panel did not consider that NSF had to move to debar the subject. The government could not continue to be in the position of providing the funds and the opportunity for these activities. Even if this case had gone to court, NSF could not expect the courts or anyone else to protect the integrity of federal research funds. NSF had to do this itself, and only the misconduct regulation makes debarment possible in a case like this. Neither "personnel or equal opportunity regulations" nor "generally applicable legal and social penalties" are adequate to safeguard the integrity of federal funds.

The special features of this case distinguish it from a common sexual harassment case. The subject used his position as a research director and mentor to create opportunities to make impermissible sexual demands and even assaults on his students and teaching assistants. The students had to submit to these demands and assaults as a condition of receiving his services as a mentor. They were particularly dependent on him not only because he gave the final grades, but also because he was the only faculty member present at the isolated research site and the other places where these events occurred. These demands and assaults, plus the professional blackmail mentioned earlier, were an integral part of the subject's performance as a research mentor and director and ethically compromised that performance. Hence, they seriously deviated from the practices accepted in the scientific community.

I have argued that certain actions are misconduct in science whether or not they are subject to other penalties and whether or not they are unique to the conduct of science. I would suggest that the appropriate criterion should be whether those actions tend to do serious harm to science. The Academy report makes a similar point: The distinguishing mark of misconduct in science, as opposed to other offenses, is that such actions "directly damage the integrity of the research process" (p. 28). The research process includes "the training and supervision of associates and students" (p. 18).

These considerations further help to explain NSF's treatment of this case. NSF would not treat a common sexual offense as misconduct in science, even if it occurred in a research setting. However, mentorship is an integral part of science, and science can be harmed by other actions besides issuing false data or stealing credit from one's peers: Science is harmed when students in a Research Experiences for Undergraduates project are taught to advance themselves by submitting to a research director's sexual demands.

OIG did not anticipate, much less seek, a case of this kind. There may be no case quite like it in the future. The important thing is that government agencies must not adopt a definition that is limited to the common run of cases so that they prevent themselves in advance from being able to deal with unexpected cases like this one. NSF's definition is "open-ended" for this reason.

OTHER EXAMPLES OF "OTHER SERIOUS DEVIATION"

Many allegations that come to OIG are about actions that do not fall under falsification, fabrication, or plagiarism. Except for the case just discussed, none of these cases has gone through the full process of investigation and adjudication. Hence, neither OIG nor NSF as a whole has made a decision that actions of these kinds would be genuine "other serious deviations." Still, among the cases OIG has received there are some strong candidates that deserve discussion. These examples again illustrate the need for a definition that goes beyond fabrication, falsification, and plagiarism.

One example of such a case involves tampering with research experiments. This would be another kind of misconduct in science that does not have to involve deception. The Academy panel criticized OIG for considering a case of this kind as misconduct in science. It reasoned that tampering is "other misconduct" rather than misconduct in science. Tampering, the report stated, is a form of vandalism or destruction of property, which again is not unique to the conduct of science and is subject to generally applicable legal and social penalties (p. 29, and p. 34, n. 20).

There is actually a range of offenses that may fall under tampering. In some instances, tampering may consist of making adjustments to a colleague's experiment without that person's knowledge, so that bad data are obtained. It is difficult to see why falsifying one's own data is misconduct in science but falsifying a colleague's data is not. In other cases, the colleague's entire apparatus may be destroyed and removed. This clearly is vandalism, but arguably there would be enough harm to research in such a situation to justify opening a case of misconduct in science. The normal penalties for vandalism would not protect the integrity of federal research funds. Similarly, a colleague's cultures may be maliciously destroyed without the destruction of any equipment. This often could not be prosecuted as vandalism, but again I think scientists would agree that it violates ethical standards and departs from accepted practices in science.

Several other examples of "other serious deviation" can be suggested. Researchers often share cultures or reagents with colleagues in other laboratories. This may be done under an exclusionary agreement—for example, an agreement that the materials not be given to a third party or that they not be used for experiments that the originator wants to perform and publish. Violation of such an agreement arguably would be misconduct in science because such violation tends to discourage a practice of sharing that is fundamental to the process of research.

Another possible "other serious deviation" is misrepresentation in grant proposals or fellowship applications. An applicant may misrepresent his or her own qualifications and achievements or may misrepresent the institution's qualifications and programs. For example, untrue claims may be made about the institution's programs in support of minority students in order to encourage favorable consideration by the agency. This again is unethical and can be a serious deviation from the standards of the scientific community.[10]

Finally, reviewers of grant proposals are instructed to keep the contents of the proposals and the opinions of other reviewers confidential. They are not supposed to use materials in the proposals for their own purposes. If these conditions are violated, harm is done to the whole process of submitting and reviewing proposals. Applicants may be afraid to write down and send in good ideas, and reviewers may feel they cannot be candid. Hence, violating the confidentiality of peer review seems to be an obvious instance of misconduct in science.[11]

HOW TO INTERPRET NSF'S DEFINITION

Because the panel misunderstood the NSF definition, it may be useful if I explain my understanding of it. Far from being a worrisome add-on, the "other serious deviation from accepted practices" phrase is central to the NSF definition. This definition says, in effect, that misconduct in science is serious deviation from accepted practices. Falsification, fabrication, and plagiarism are mentioned as outstanding examples. Then the definition goes on to say that all "other" actions that similarly deviate from accepted practices are also misconduct in science.

However, I suggest that NSF, unlike the Academy panel, understands "deviation from accepted practices" in an ethical sense. The way to commit misconduct in science is to do something that scientists would recognize as deviating seriously from professional ethical standards. The panel evidently took "accepted practices" to mean accepted ways of doing experiments. Deviating from those does not ordinarily involve any ethical violation and has nothing to do with misconduct. Those who drafted the NSF definition obviously did not contemplate an interpretation that would make it misconduct in science just to do something novel or unorthodox.[12]

The NSF definition does not attempt to give a full list of the practices that would violate professional standards in science. It might be very hard to draw up an exhaustive list, and standards might be found to vary from one branch of science to another. By referring to "accepted practices," the NSF definition points to the relevant scientific community as the authority for what is or is not misconduct. Such a definition is heuristic rather than vague. It does not say whether each and every practice is or is not misconduct, but it points out where to look for the answer. The assumption is that working scientists, like the members of other professions, can and ought to know the standards of their profession and that in disputed cases representatives of the scientific community can agree on what those standards are. They should be able to do this without being given a complete list of the types of misconduct. Hence, misconduct can be recognized and dealt with under a heuristic definition like NSF's.

Some scientists may be willing to have their academic colleagues deal with misconduct in this heuristic way but may be less comfortable about a government agency doing so. This is not the place to discuss NSF's competence or public distrust of government. So far, no case has gone to adjudication at NSF that involved disagreement over whether an alleged activity would be misconduct in science. If that were to happen, I expect that a satisfactory method of consultation between the agency and the scientific community could and would be worked out.

CONCLUSION

NSF uses an open-ended definition that contains the phrase "other serious deviation from accepted practices." To date, this definition has worked successfully. One of its major advantages is that it leaves the agency the possibility of taking action when a case arises that is not on some short list of types of misconduct. It is legitimate to ask how NSF understands this definition, how it was applied in a major case, and what safeguards there are against abuse. If the Academy panel had asked, it might have produced more helpful recommendations and might have advanced the discussion of this subject much more than it did. Those who work on misconduct cases will always need the guidance and insights of their colleagues in the broader scientific community. But those who wish to make useful policy recommendations also need the insights of those with day-to-day experience in this highly controversial area.

Notes

1. National Academy of Sciences, Panel on Scientific Responsibility and the Conduct of Research, Committee on Science, Engineering, and Public Policy, *Responsible Science: Ensuring the Integrity of the Research Process,* Vol. 1 (Washington, D.C.: National Academy Press, 1992).

2. The Public Health Service definition is that "Misconduct" or "Misconduct in Science" means fabrication, falsification, plagiarism, or other practices that seriously deviate from those that are commonly accepted within the scientific community for proposing, conducting, or reporting research. It does not include honest error or honest differences in interpretations or judgments of data (45 C.F.R. sec. 50.102). The NSF definition is that "Misconduct" means (1) fabrication, plagiarism, or other serious deviation from accepted practices in proposing, carrying out, or reporting results from activities funded by NSF or (2) retaliation of any kind against a person who reported or provided information about suspected or alleged misconduct and who has not acted in bad faith [45 C.F.R. sec. 689.1(a)].

3. On the other hand, some panel members—a majority, by some accounts—dissented and wanted to include other actions as misconduct under the Academy's definition. The report lists as examples "misuse of the peer-review system to penalize competitors, deceptive selection of data or statistical analysis, or encouragement of trainees to practice misconduct in science" (p. 27). It goes on to say that "These issues deserve further consideration by the scientific research community to determine whether the panel's definition of misconduct in science is flexible enough to include all or most actions that directly damage the integrity of the research process and that were undertaken with the intent to deceive" (p. 28). Since this further discussion has not yet taken place, it seems premature to urge that agencies limit their definition to the three items on which all panel members could agree. By doing this, these agencies would preclude themselves from dealing with other situations, such as the ones just listed.

4. OIG receives complaints of many kinds that may involve various laws or regulations in various combinations. Some complaints are clearly about misconduct in science, but many are harder to classify. Most cases are concluded informally, without full investigation and adjudication, so that there may never be a decision about whether misconduct in science was involved. Universities also may resolve complaints against their students or faculty members without making it clear whether they are employing their misconduct-in-science regulations or some other disciplinary procedure. The result is that OIG's files contain some cases, including closed cases, that are not definitely classified as to whether they involve misconduct in science or not. Furthermore, the closed cases tend to be ones that were resolved quickly because it was easy to show that there was no misconduct. Some serious cases that require full investigation could not be closed quickly and remain open. Therefore, it is premature at this

stage to draw any conclusions about what proportion of our cases will eventually lead to findings of misconduct or about what the distribution of those cases by type of misconduct will eventually be.

5. In fact, 131 were received, but 7 have been reclassified and transferred to other offices within OIG.

6. These cases are discussed more fully in OIG's series of semiannual reports to the Congress, which are available to the public. See, for example, OIG, *Semiannual Report* to the Congress, NSF, numbers 3 through 7.

7. This was the view of the Academy panel (pp. 28–29:3). It was also apparently the view of the Public Health Service Advisory Committee on Scientific Integrity, which recommended the following definition at its June 1992 meeting: "Research fraud is defined as plagiarism, fabrication or intentional falsification of data, research procedures, or data analysis; or other deliberate misrepresentation in proposing, conducting, reporting or reviewing research" (Minutes of the Meeting of the Department of Health and Human Services, Public Health Service Advisory Committee on Scientific Integrity, 11–12 June 1992).

8. The available published information was too limited to permit a serious evaluation of this case. Under the Freedom of Information Act, the staff of the Academy panel requested and received a copy of OIG's investigation report, but I have not found any panel member who saw it. In any case the arguments in the OIG investigation report are not addressed in the panel's report. The staff and panel also did not discuss the case with NSF.

9. A further argument, that these behaviors "should be governed by mechanisms that apply to all institutional members, not just those who received government research awards" (p. 86), obviously applies to a grantee institution and not to a federal agency.

10. The fabrication of bibliographic material in research proposals is mentioned in the report, and the claim is made that the proposed definition covers it (p. 86). However, the proposed definition speaks only of the fabrication of data of results (pp. 27 and 47).

11. The Academy report characterizes the misuse of privileged information as plagiarism, which it defines as using the words or ideas of another person without giving appropriate credit (pp. 54–55). However, using privileged material without authorization is much different from using published material and not giving credit. The privileged material is not supposed to be used at all, even if credit is given. NSF grant applicants, for instance, do not expect the ideas in the proposals to be used by everyone who is willing to give an acknowledgment. Hence, plagiarism is not a broad enough notion to cover this kind of misconduct.

12. An explanation of "other serious deviation" has been given by R. M. Anderson, one of the principal drafters of the NSF definition in *J. Law Techno* 3(121)(1988):129–31.

What Is Misconduct in Science?

HOWARD K. SCHACHMAN

Answering the question posed in the title depends on one's perspective. One could focus on collegial behavior in an academic setting. In that case the discussion would be far-ranging and might include attempts to formulate ethical principles and guidelines for the conduct of research. It would examine complex problems involving the sharing of data and unusual materials, as well as authorship and publication practices. It would certainly include condemnation of egregious actions such as plagiarism and the fabrication and falsification of data and results. From such an examination one could formulate a definition of misconduct in science that would form the basis for governmental action leading potentially to debarment from federal support. Such a sanction, in effect, could lead to the termination of a career in science. For such an outcome a precise, rigorous, and unambiguous definition of misconduct in science is essential. Governmental oversight over the expenditure of taxpayers' money is legally mandated and clearly proper. It is obligatory for the National Science Foundation (NSF) and National Institutes of Health (NIH) to investigate allegations of fraudulent acts and to impose sanctions when guilt is demonstrated. In contrast, it is inappropriate, wasteful, and likely to be destructive to science for government agencies to delve into the styles of scientists and their behavioral patterns.

The definitions of misconduct in science currently used by governmental agencies unfortunately intermix these two different aims.[1] In defining misconduct as fabrication, falsification, and plagiarism, NSF and NIH also include an open-ended phrase to encompass "other seri-

Reprinted with permission from *Science* 261 (1993):148–49, 183, and by permission of the author.

The author is in the Department of Molecular and Cell Biology, University of California, Berkeley, Calif.

ous deviation from accepted practices in proposing, carrying out and reporting results." Because these definitions are overly broad and vague, it is appropriate to examine the history of congressional investigations of fraud in research and to consider a definition that is consistent with and responsive to the intent of Congress in establishing oversight of federal funds for scientific research.

Many scientists, like others in our society, are ambitious, self-serving, opportunistic, selfish, competitive, contentious, aggressive, and arrogant; but that does not mean they are crooks. It is essential to distinguish between research fraud on the one hand and irritating and careless behavioral patterns of scientists, no matter how objectionable, on the other. We must distinguish between the crooks and the jerks.[2] For the former we need (i) governmental oversight, (ii) a clear definition of those acts that are proscribed, (iii) adjudicatory machinery, (iv) due process, (v) protection of whistle-blowers, (vi) strong sanctions for the guilty, and (vii) full disclosure of conclusions in order to minimize repetition in other institutions. In contrast, such governmental intervention is inappropriate for concerns regarding errors in collecting and interpreting data, incompetence, poor laboratory procedures, selection of data, authorship practices, and multiple publications. These are matters for explicit dialog and education in universities and research institutions.

If we are to avoid the imposition of guidelines, rules, and regulations that may impede scientific research, it is essential to limit governmental action to fraud in science. A definition of misconduct in science that recognizes the dichotomy of roles and the need to "render, therefore, unto Caesar the things which are Caesar's" will reduce the tension now existing between working scientists and government officials.

HOW "FRAUD IN SCIENCE" BECAME "MISCONDUCT IN SCIENCE"

In 1981 a subcommittee of Congress, under the chairmanship of Congressman Albert Gore, Jr., held hearings on fraud in biomedical research[3] in response to widespread reports of scientists falsifying their data. The cases cited dealt with fraud and plagiarism, One witness described how he falsified results of experiments that had not been performed. Another case, as described by the chairman, involved a researcher who "became entangled in a network of fraud and plagiarism, and a possible cover-up." Throughout these hearings the focus was on fabrication, falsification, and plagiarism.

When Congress passed the Health Research Extension Act in 1985, the legislation directed the secretary of the Department of Health and Human Services to require institutional applicants for NIH funds to review reports of fraud and report to the Secretary any investigation of suspected fraud which appears substantial.

The language focused on fraud, and the director of NIH was required to establish "a process for the prompt and appropriate response to information provided the Director . . . respecting scientific fraud."

Several years later, following increasing media coverage of several notorious cases of fabrication and falsification of data, the language was altered significantly when the Public Health Service (PHS) issued a proposed rule[4] entitled "Responsibilities of PHS Awardee and Applicant Institutions for Dealing with and Reporting Possible Misconduct in Science." In that proposed rule, "misconduct in science" was defined as (i) fabrication, falsification, plagiarism, deception or other practices that seriously deviate from those that are commonly accepted within the scientific community for proposing, conducting or reporting research; or (ii) material failure to comply with Federal requirements that uniquely relate to the conduct of research.

Meanwhile, NSF issued final regulations under the title "Misconduct in Science and Engineering Education" that defined misconduct and also provided a safeguard for reprisals against whistle-blowers.[5]

It was this transition from "fraud in science" to "misconduct in science" that led to apprehension among scientists. Some of the actions described in congressional hearings are labeled appropriately as fraud. Faking data is fraudulent. So is falsifying data. There is little confusion over the meaning of fraud. In contrast, "misconduct in science" means different things to different people. The change to "misconduct" instead of "fraud" was initiated and effected by lawyers and not by scientists. It was because of the legal burden of having to prove intent and injury to persons relying on fraudulent research that counsels for NSF and PHS wanted the change to misconduct.[6] My concern is over vagueness of the term "misconduct in science" and how people with different orientations interpret various alleged abuses.

In formulations of the term "misconduct in science" there is agreement on fabrication, falsification, and plagiarism. Scientists have emphasized that "misconduct in science" does not include factors intrinsic to the process of science, such as error, conflicts in data, or differences in interpretation or judgments of data or experimental design.[7] Particularly bothersome was inclusion of the phrase "other practices that seriously deviate from those that are commonly accepted within the scientific community for proposing, conducting or reporting research."

Not only is this language vague but it invites over-expansive interpretation. Also, its inclusion could discourage unorthodox, highly innovative approaches that lead to major advances in science. Brilliant, creative, pioneering research often deviates from that commonly accepted within the scientific community.

My apprehension over this open-ended, vague section of the definition is best illustrated by a case cited by the Office of Inspector General (OIG) of the NSF:[8]

> In November, 1989, OIG received allegations of misconduct against the researcher. Our investigation involved conducting extensive interviews and collecting affidavits . . .

OIG determined that the researcher had been involved in 16 incidents of sexual misfeasance with female graduate and undergraduate students at the research site; on the way to the site; and in his home, car, and office. Many of these incidents were classifiable as sexual assaults. OIG further determined that these incidents were an integral part of this individual's performance as a researcher and research mentor and represented a serious deviation from accepted research practices. Therefore, they amounted to research misconduct under NSF regulations.

This is a preposterous and appalling application of the definition of scientific misconduct. The individual involved in this case, assuming the allegations were proven, should have been terminated by his institution for moral turpitude and the grant canceled accordingly. All of the grant funds should have been returned to the government by the institution that employed the individual. This case is an example of misconduct for which institutional and legal sanctions should have been imposed. But it is not misconduct in science. Having read the investigative report on this case, I am convinced that charges of sexual harassment as well as sexual abuse should have been filed. Buzzelli,[1] however, reached an opposite conclusion, stating that "This case was not essentially a sexual harassment case, but sexual offenses were obviously at the heart of it . . ."

DEFINING MISCONDUCT IN SCIENCE

In 1992 a panel convened by the National Academy of Sciences (NAS), National Academy of Engineering, and the Institute of Medicine released a report[9] that defined misconduct in science as fabrication, falsification, or plagiarism, in proposing, performing, or reporting research. Misconduct in science does not include errors of judgment; errors in the recording, selection, or analysis of data; differences in opinions involving the interpretation of data; or misconduct unrelated to the research process.

Fabrication is making up data or results. Falsification is changing data or results. Whereas plagiarism is described in the report as "using the ideas or words of another person without giving appropriate credit," Webster's *Seventh New Collegiate Dictionary* defines "plagiarize" as follows: "to steal and pass off as one's own (the ideas or words of another); to present as one's own an idea or product derived from an existing source." Because of the increasing focus on "intellectual property" in recent years, plagiarism is best defined as "misappropriation of intellectual property." Defined in this way, plagiarism not only encompasses those cases in which sentences or phrases are used without attribution but also includes unauthorized use of ideas, data, and interpretations obtained during the course of the grant review process or the review of scientific papers being considered for publication.[10]

It is fabrication, falsification, and plagiarism that attracted the attention of the congressional committees, chaired by former Congressman Gore, Congressman John Dingell, and the late Congressman Ted Weiss.[11] In the two most publicized cases that have dominated news disparaging the scientific community in the past few years, the initial charges were focused on these matters. Was the virus misappropriated? If so, a verdict of misconduct in science is correct. In the other case, it is important to know whether the experiments were done. If they were not, a verdict of misconduct is appropriate. One need not have a vague, open-ended phrase in the definition to adjudicate these cases. Reaching a verdict on grounds of fabrication, falsification, or plagiarism is difficult enough; there is no need to make the adjudication even more complex by considering spurious or vague charges as well.

RISKS OF AN OPEN-ENDED DEFINITION

Those who advocate the desirability of the clause "other serious deviation" have presented a variety of scenarios.[1] One is tampering with research experiments. This, like sabotaging experiments and destroying animal quarters, is covered by other statutes and is, and should be, subject to sanctions. But we must face the fact that NSF could not impose sanctions on an individual who does not have an NSF grant even though that person tampers with or sabotages an experiment of an individual supported by NSF. Clearly, including such cases as misconduct in science leads to a morass. These are problems for local institutions and statutes dealing with vandalism. Invoking the "seriously deviates" clause to impose sanctions for such actions and labeling them misconduct in science is a great mistake.[12]

Other examples, such as misrepresentations of one's qualifications and achievements in a grant application, are covered by falsification. The clause "seriously deviates" is also applied to reviewers of grant proposals who violate confidentiality and use materials in the proposals for their own purposes. This doubtless happens, and the cases should be investigated. If guilt is established, sanctions should be imposed. But one does not need an open-ended, vague, unclear phrase in the definition to encompass such egregious behavior. It is amply described as misappropriation of intellectual property and, therefore, encompassed in the definition as plagiarism.

The inclusion of ambiguous terms in the definition of misconduct in science potentially breaches an important principle of due process, the right to know in advance those activities that are proscribed. This principle is certainly violated by the view that ". . . you have to have a definition that covers situations that you can't even now conceive of."[13]

Although the word "misconduct" is now used in order to avoid legal ramifications of the word "fraud," it is nonetheless important to retain the original intent of Congress to focus on the role of government in investigating misconduct in science that is equivalent to "fraud which appears substantial." It is encouraging that the PHS Advisory Committee on Scientific Integrity has recently recommended a major change in the definition of misconduct in science now being used by the Office of Research Integrity. This proposal eliminates the phrase "other practices that seriously deviate from those that are commonly accepted within the scientific community" and moves closer to that proposed by the NAS panel.[9,14] Also, the PHS will no longer list in its ALERT system those individuals under investigation. This terrible practice of including names of individuals under investigation for misconduct in science has been abandoned; now names will be listed only if a finding of guilty has been reached. History is full of examples of governmental promulgations of laws expressed in broad, open-ended terms that were elastic enough to be stretched to cover any individual action that irritated some officials. In this century alone it was a major offense in some countries to publish scientific papers that seriously deviated from accepted practice. The enforcement of such strictures virtually destroyed major areas of science in those countries. We should not expose science in this country to similar risks.

Notes

1. D. E. Buzzelli, *Science* 259(1993):584.

2. C. K. Gunsalus, paper presented at the Annual Meeting of the American Association for the Advancement of Science, Chicago, 6 February 1992, Symposium on Integrity and Misconduct in Science.

3. Hearings before the Subcommittee on Investigations and Oversight of the Committee on Science and Technology, U.S. House of Representatives, 97th Congress, 31 March to 1 April 1981.

4. Section 493 of the amended Public Health Service Act constitutes the Enabling Act requiring the secretary of Health and Human Services to issue regulations requiring investigation of "alleged scientific fraud which appears substantial." This language is especially significant in terms of the wide variety of misdeeds now subject to investigation and imposition of sanctions by NSF and NIH. K. D. Hansen and B. C. Hansen (*FASB J.* 5(1991):2512), in their critical analysis of "Scientific Fraud and the Public Health Service Act," emphasized that the amendment is clear on its face but there has been a tendency by the agencies to greatly expand the authority granted.

5. The current NSF definition of misconduct in science is (i) fabrication, falsification, plagiarism, or other serious deviation from accepted practices in proposing, carrying out, or reporting results from activities funded by NSF, or (ii) retaliation of any kind against a person who reported or provided information about suspected or alleged misconduct and who has not acted in bad faith (45 C.F.R. sec. 689).

6. R. M. Anderson, *Select Legal Provisions Regulating Scientific Misconduct in Federally Supported Research Papers,* AAAS-American Bar Association National Conference of Lawyers and Scientists Project on Scientific Fraud and Misconduct, Report on Workshop Number 3, (Washington, D.C.: American Association for the Advancement of Science, 1989).

7. Testimony by H. K. Schachman before the Subcommittee on Investigations and Oversight of the Committee on Science, Space, and Technology, U.S. House of Representatives, 101st Congress, 28 June 1989.

8. Semiannual Report of the Office of Inspector General of the National Science Foundation, 1 April to 30 September 1990.

9. *Responsible Science: Ensuring the Integrity of the Research Process,* Vol. 1 (Washington, D.C.: National Academy Press, 1992).

10. Many of the cases of misconduct in science are described as plagiarism (First Annual Report of Scientific Misconduct Investigations Reviewed by Office of Scientific Integrity Review, March 1989 to December 1990, of the Public Health Service; Semiannual Report of the Office of Inspector General of the

National Science Foundation, No. 6, 1 October 1991 to 31 March 1992, and No. 7, 1 April 1992 to 30 September 1992). A definition of plagiarism as misappropriation of intellectual property would suffice for adjudicating the case at Michigan State University reported by E. Marshall (*Science* 259(1993):592). Based on the findings described by the independent panel (as reported in *Science*), the verdict of misconduct in science should have been attributed to plagiarism. There is no need to invoke the clause "a serious deviation from accepted practices."

11. In addition to the hearings in note 3, a subcommittee of the Committee on Government Operations, House of Representatives, 100th Congress, held hearings on 11 and 12 April 1988 dealing with "Scientific Fraud and Misconduct and the Federal Response" under the chairmanship of the late Congressman Ted Weiss. On 12 April 1988, a hearing on "Fraud in NIH Grant Programs" was held by the Subcommittee on Oversight and Investigations of the Committee on Energy and Commerce, House of Representatives, 100th Congress, under the chairmanship of Congressman John D. Dingell. The Subcommittee on Human Resources and Intergovernmental Relations of the Committee on Government Operations of the U.S. House of Representatives, 100th Congress, held hearings on 29 September 1989 entitled "Federal Response to Misconduct in Science: Are Conflicts of Interest Hazardous to Our Health?" "Scientific Fraud" was the title of hearings of the Subcommittee on Oversight and Investigations of the Committee on Energy and Commerce of the U.S. House of Representatives, 101st Congress, 4 to 9 May 1989 and 30 April and 14 May 1990.

12. One might wonder whether a scientist who uses NSF funds to employ an illegal alien as a technician will be guilty of misconduct in science rather than of violating immigration and, perhaps, tax laws.

13. D. P. Hamilton,"OSI: Better the devil you know?" *Science* 255(1992): 1345.

14. Eliminating this open-ended part of the definition will reduce the burdens on governmental officials, thereby facilitating their concentration on fraud of a substantial nature. Already the staff at ORI numbers about 50 people with an annual budget of $5 million.

8

Animal Experimentation and Ethics

DENI ELLIOTT and MARILYN BROWN

I. INTRODUCTION

Why discuss ethics in animal experimentation? Why are animals used in research, testing, and teaching? What is all the controversy about? Is there a difference between animal rights and animal welfare? Are there laws governing how animals are used? What other responsibilities do scientists have when they use animals in research, teaching, and testing?

There is some kind of justification necessary if a researcher is causing suffering, distress, or death to nonhuman animals in the process of research, but is that because it is a moral problem? Proponents and opponents of animal use alike refer to the similarities or differences between human and nonhuman animals as a foundation from which to make their case. It is possible to argue for judicious and limited use of animals in research without engaging in debates like this. Regardless of how nonhuman animals are like or not like humans, we have a long history of protecting them from harm, along with our long history of hunting them, of raising them for food, and of using them for research purposes. It makes sense to consider animal use for research purposes in this broader sense.

II. A SHORT HISTORY OF REGULATION RELATING TO ANIMAL USE

The primary regulations governing the use of animals in research, testing and teaching are the Animal Welfare Act (AWA) and the Public Health Service Policy on the Humane Care and Use of Laboratory Animals (PHS Policy), which uses the *Guide for the Care and Use of Laboratory Animals* as the "yardstick" by which animal care and use programs are evaluated.

The AWA was passed in 1966. It was revised in 1970, 1976, and 1985. Original versions of this regulation primarily set standards for animal care (such as cage size, sanitation, feeding and watering); however, over the years the AWA has evolved to also impact on how animals are used. Significant changes include greater details on the responsibilities of the institution/facility, the responsibilities of the veterinarian and the components of a program of adequate veterinary care, and the functions and responsibilities of the Institutional Animal Care and Use Committee (IACUC).

The IACUC membership requirements are slightly different in these two regulations; however, both require that IACUCs at least include the attending veterinarian, a scientist, and an unaffiliated member to represent the community's interest and concerns. This committee is responsible for evaluating all animal protocols and procedures using a list of criteria including such items as justification of animals used and numbers used; level of potential pain and distress and how it will be minimized using appropriate anesthetics, analgesics, and tranquilizers; surgical and postsurgical care; qualifications of personnel using the animals; methods of euthanasia; and documentation that alternatives to potentially painful procedures have been adequately considered. The IACUC has the authority, indeed the responsibility, to suspend any projects that are not in accordance with an IACUC-approved protocol. The IACUC also conducts semiannual evaluations of the animal care and use program, including inspecting facilities and reviewing concerns presented regarding animal welfare.

In addition to the regulatory requirements, the use of animals may be subject to rules from other federal agencies such as the Food and Drug Administration or private funding organizations. Many institutions using animals are also involved in the voluntary accreditation process of the American Association for the Accreditation of Laboratory Animal Care.

Regardless of which laws and regulations apply, the public holds the scientific community responsible and accountable for the judicious and humane care and use of laboratory animals.

III. THE MORAL COMMUNITY MISUNDERSTOOD

The question of what humans owe to nonhuman animals and why they would have such obligation has traditionally been argued over the boundary of what philosophers call the "moral community." The community, which lists those who are deserving of equal moral protections, has become the battlefield upon which those in favor of animal use and those opposed have fought the theoretical question.

Those who argue that animals *should* be used in research often base their arguments on the notion that a human life is worth more than the life of a mouse or that of a research rabbit, dog, or monkey. Animals are believed to lack the standing to have moral protections equal to humans. For example, Machan argues his case this way:

> Normal human life involves moral tasks and that is why we are more important than other beings in nature—we are subject to moral appraisal, it is a matter of our doing whether we succeed or fail in our lives.[1]

Those who argue that nonhuman animals *should not* be used for research argue that animals have claim to moral protections that are equal in status to those of humans. Sapontzis argues it this way:

> [W]hile it is true that humans are ordinarily capable of flights of reason of which animals are not, it does not immediately follow either that we are morally superior to animals or that we are morally justified in sacrificing the interests of less rational beings for our benefit.[2]

People disagree about how to determine "worthiness" for moral consideration. A being within the moral community has a fundamental right to moral consideration: A purposeful action against that being that is likely to cause harm is a prima facie wrong. In the end, arguments for or against inclusion of animals in the moral community may suffer from being overly inclusive or under inclusive, or both.

The scope of the moral community is overly inclusive when those arguing in favor of including nonhuman animals are no less arbitrary in drawing the line between what to include and what to exclude than are those who insist that the line that exists between human and nonhuman animals. On what basis do you include baboons but not bats? Bats but not beetles? Beetles but not bacteria?

On the other hand, arguments are underinclusive if they base inclusion in the moral community on the rationality of its members. In doing so, these arguments to keep the moral community as an exclusively human domain end up excluding a number of humans: infants, children, people with developmental disabilities, and those incapacitated by illness or advanced age.

Occasionally, the criterion for membership is simultaneously over- and underinclusive. For example, Singer says,

> If a being suffers, there can be no moral justification for refusing to take that suffering into consideration. No matter what the nature of the being, the principle of equality requires that its suffering be counted equally with the like suffering—in so far as rough comparisons can be made—of any other being. If a being is not capable of suffering, or of experiencing enjoyment or happiness, there is nothing to be taken into account.[3]

In Singer's attempt to enlarge the moral community to include animals (beings that suffer), he has managed to narrow the community to exclude humans whom we think are deserving of moral protections. Some humans are "not capable of suffering or of experiencing enjoyment or happiness," in acute or chronic periods of lost consciousness, yet this does not provide license to violate their bodies.

An alternative definition of moral community rests not on the rational capacity or the moral agency of individual members, nor does it include everything that might respond to stimulus.

According to this definition of the moral community, developed by philosopher Bernard Gert, the hallmark of membership in the moral community is that every individual member is (theoretically) deserving of moral protections equal to every other member. No one should cause any member of the moral community to suffer evils (death, pain, disability, deprivation of pleasure or freedom) without justification. The moral community includes at its core rational adults—human beings capable of deciding to whom it is theoretically possible to provide these protections—but it expands from there.

Says Gert:

> No one disagrees that one must include all presently existing moral agents; those who disagree claim that the group must include more than this. The smallest change is to claim that it must include all who were ever moral agents and remain persons, that is, are still capable of any conscious awareness. This change is supported by noting that all rational persons who are moral agents would want to retain the protection of the moral rules [Don't kill, cause pain or disability, or deprive of freedom or pleasure] if they were to lose their capacity to act as moral agents; at least they would want this protection as long as they could suffer from losing it . . . Most readers of this book would want to include in the group toward which we should impartially follow the moral rules human infants who have not yet become moral agents.[4]

Within Gert's theory, everyone in the moral community has an equal entitlement to protection. It is not moral to take the heart of one living child to save another, whatever the differences in wealth or culture. The child who needs a heart transplant and the child who has a healthy matching heart are of equal moral worth. They deserve equal moral protections.

The minimal group members of the moral community are, indeed, rational agents, but that does not imply that they have greater moral protections. The minimal group, composed of rational members who can think through the consequences of broadening membership in the moral community, are the deciders for who else belongs.

Membership in the moral community has changed as the minimal group has become more cognizant of others who are morally analogous. Christopher Stone notes, "As late as the Patria Potestas of the Romans, the father had jus vitae necisque—the power of life and death—over his children . . . The child was less than a person: an object, a thing."[5] Racial and ethnic minorities, people with disabilities, women, and children have all struggled to attain membership in the moral community that was limited, in Aristotle's time, to free adult men.

Recognizing that all born humans deserve equal moral protections does not imply that it is always immoral to cause their death, pain, or disability, or their loss of freedom or pleasure. Rather, it means that, morally speaking, justification must be provided if these evils are caused. An example of current struggle over whether causing death to a member of the moral community is justified is the social debate over whether the certain death of anencephalics soon after birth provides adequate justification for hastening their deaths.[6]

While we know that infanticide of female children occurs in some parts of the world, we also know that such killing of babies is wrong. A baby's sex is not adequate justification for causing her death. But while infanticide is morally wrong, it is only one of many inequalities resulting in suffering, distress or death that is tolerated due to social or economic circumstances. However, economic and social inequalities do not provide moral justification for not providing all members of the moral community equal moral protections.[7]

Nor is there complete agreement about who to include in the moral community. Some—people in persistive vegetative states and humans who have been conceived but not fully gestated, for example—are borderline cases. They are borderline because reasonable people (in the minimal group) disagree as to whether or not to include these kinds of beings as deserving of equal moral protections. If a fetus or a permanently unconscious (and never-again-to-be-sentient) person belongs in the moral community, it follows from definition that that being has moral protections equal to those of every other member. While it is possible to provide equal moral protections to these borderline groups, reasonable people disagree as to whether they *should* be provided such protections.[8]

Membership in the moral community is an either/or proposition. If one begins to think in terms of degrees of membership, the idea of equal moral protections is lost. In theory, anything can be considered for membership in the moral community. The necessary condition for membership is that members of the minimal group (rational moral agents) are willing to hold others in the community accountable for acting impartially toward the new members. The new members must be treated as deserving of moral protections equal to those in the minimal group. It is this condition that makes it logically impossible to include nonhuman animals in that community.

Consider what we do for members of the moral community. We construct governments to manage many of our moral protections. The general function of government is to prevent the suffering of unnecessary evils by that segment of the moral community who are the citizens.[9] We enact laws that hold people accountable for causing some kinds of harm to those within the community. We construct laws that include safety codes to prevent some harms from being caused. Many nation-states, including ours, have enacted laws that provide universal police and fire protection. Laws that protect members' lives and property extend to all members of the moral community who are under the jurisdiction of the government constructed to manage those protections.

Consider what would happen if we extended the moral community to include other mammals. As members of the moral community, they would be deserving of the protections afforded all other members. Animals in the wild would be as deserving of moral protections as those in the cities. If we advocate protecting some members against predators, we must advocate protecting all members (even if we know that sometimes members get hurt anyway). We would need to begin with the premise that nonhuman mammals are deserving of protections that we give all other members from humans as well as from other animals that might consider them food. Government officials do not hesitate to hunt down and kill predators such as mountain lion and bear when they treat human beings as prey.

Practically speaking, of course, this is impossible. Humans can survive without eating meat, but the cougar cannot. Our intention of protecting individual nonhuman animals would have the unintended result of the destruction of species that depend on the flesh of other mammals. Or equal protection would dictate extensive factory farming of animals considered acceptable to be used as "food" for nonhuman animals, just as we currently factory farm for human use.

We could get around that problem by allowing into the moral community only omnivorous or herbivorous species. But that creates a standard that, on its face, seems even more arbitrary than the distinction between those now included or excluded.

The only way to include animals in the moral community is to make "moral community" mean something other than deserving of equal moral protections. If one does that, "moral community" loses its power to help us deal with a broad set of problems. Knowing who is entitled to equal moral protections is important in determining what counts as morally prohibited activity and what counts as activity requiring justification.

That everything in the moral community deserves equal moral protections does not imply that *only* things within the community are deserving of *any* protections. There are things outside the moral community (including research animals) that deserve some moral protections, but those protections are less than those given to members of the moral community and are justified on a different basis.

IV. ALL MORAL STANDING ASIDE

All moral standing aside, a case can be made for the protection of animals. The present argument is confined to the use of animals in research, but the argument implies restrictions against animal death, suffering, and distress that occur in other activities such as recreational hunting, factory farming, and pet ownership. The discussion focuses on the use of animals in laboratory experiments that result in their death or in their suffering or distress.

The discussion does not include what counts as allowable levels of animal "pain." There is disagreement about how to interpret animal pain. One author presents "animal pain" as an oxymoron:

> Pain is the body's representative in the mind's decision-making process. Without pain, the mind would imperil the body . . . But without the rational decision-making mind, pain is superfluous. Animals have no rational or moral considerations which might overrule the needs of the body.[10]

However, other scientists point out that clinical pain studies rest on the assumption that animal pain is analogous to human pain.[11]

However one interprets animal pain, there seems to be more agreement that at least some animals suffer or become distressed. Bernard Rollin, who is a physiologist and biophysicist as well as a professor of philosophy, explains the reason for focus on the animal's interpretation of experience:

> Recent research has shown, for example, that talking about stress in animals cannot be done in purely mechanical terms and cannot circumvent reference to their states of awareness, for the same "stressor" can have very different physiological effects in an animal, depending on the animal's emotional state, or how it has been treated prior to the noxious stimulus, or on whether it can anticipate or control the stimulus, etc. Furthermore, purely psychological stressors, like putting an animal in an unfamiliar environment, can have greater physiological effects than such physical stressors as heat.[12]

This is relevant to the argument for the protection of research animals. If the research scientist has reason to believe that animals in her care do suffer or become distressed, then that knowledge (reason to believe) alone creates some moral obligation on the part of the research scientist.

Argument 1: Based on the Perceptions of Those in the Moral Community

One of the moral protections afforded those in the moral community is that when harm is caused, it must be justified.

Some people are hurt by the knowledge that animals are caused suffering and distress. Some people do not want animals to be caused suffering and distress at all. An even larger number of people do not want animals caused suffering and distress without justification. If research scientists are causing pain to these clearly human members of the moral community (by causing animal death, suffering, or distress), they must justify those acts.

The operable term here cannot be "knowledge." That is, one can't argue against this need for justification by claiming that what people don't know won't hurt them. Abusing a child, for example, is wrong even if the authorities are not aware of the situation.

It's true that there are many things that are not noticed as moral problems unless people have knowledge of the act in question. But it doesn't follow that actions are outside the bounds of moral consideration if they take place in secret. People who object to the unjustified causing of death, suffering, and distress to research animals object to that action, not to the knowledge of it.

If the use of research animals should be limited because people object to unlimited use, then, the justification for when animals are used must be satisfactory to those who object. Adequate justification, not surprisingly, is a utilitarian argument that cites the morally relevant difference between those who are members of the moral community and those who are not.

The moral justification for causing death, suffering, and distress to animals is that human lives, and additionally, animal lives, are being saved or significantly improved through the animal use. Now, there is argument about when this justification is valid. For example, one investigator uses this balancing as a way of justifying pain studies in rats:

> No investigator views the burning procedure with equanimity. However, the realization that research in animals with burns may help to reduce the suffering and enhance the recovery of young human burn victims well justifies this work.[13]

Even those who argue against animal use cite this utilitarian balancing. Says Sapontzis, "we consume hundreds of millions of laboratory animals each year. It is hard to imagine that our testing and research produces utilitarian goods sufficient to outweigh that massive, annual evil."[14]

Consider the following scenario: Three day-old chickens are killed, with the result that hundreds of human children are protected against some childhood disease. The chicks are killed in a way that does not cause them to suffer or become distressed prior to their death.

It's admittedly rare in research that the cost and benefit equation is that clean and direct, but the fact that this justification would satisfy those members of the moral community who approve of animal use if justified moves the problem from the moral to the empirical realm. Now the question about whether specific animal use is justified depends on how close the actual intended use is to the chick-to-children scenario.

Argument 2: From the Sentience and Perception of the Animal

Some nonhuman animals are part of the class of a small number of things outside of the moral community that we know to suffer and become distressed. Whether we refer to the stewardship role implicit in what it means to be "humane" or to our sense of empathy, it is the case that we think there is something wrong with causing unnecessary death, suffering, and distress to animals. In our society, it is both a crime and a psychiatric illness.[15]

The knowledge that one can cause suffering and distress to something outside of the moral community carries with it the moral responsibility to avoid or minimize that suffering and distress when possible. The difference between speculation that vegetables may suffer when cut from the vine and that animals suffer from a surgical incision if inadequately anesthetized is that we know, with certainty, that the latter is true. Knowledge that certain actions cause suffering and distress implies moral obligation to avoid or minimize that suffering and distress.

As Rollins explains,

> Most realize that as soon as one has admitted that animals can be hurt in ways which matter to them . . . or that unnecessary animal suffering is wrong, one has implicitly but inescapably presupposed that animals are in the moral arena, that one can be morally wrong in how one uses or treats animals, none of which one would say of inanimate objects, such as chairs and wheelbarrows.[16]

Argument 3: From the Special Role of the Research Scientist

The goal of research science, broadly speaking, is to seek and communicate new understandings about nature. Indeed, we're told that "the desire to observe or understand what no one has ever observed or understood before is one of the forces that keeps researchers rooted to their laboratory benches, climbing through the dense undergrowth of a sweltering jungle, or pursuing the threads of a difficult theoretical problem."[17] Scientists act using methods that can, in principle, be replicated.

Animals are first a part of nature that is explored by science. Research scientists have a special responsibility for stewardship—judicious care—over the realm in which they work. In structure, this stewardship responsibility is no different from claiming that journalists have a special responsibility to protect the First Amendment.

There's something ironic about a geologist who destroys rock formations without reason or a botanist who, through neglect, allows his or her plants to die. The researcher's role as student of natural phenomena implies a responsibility to care for and respect the objects of study. Specifically, in the case of animal research, scientists have an obligation to prevent the death, suffering, and distress of the animals that are their objects of study whenever possible. Animals ought not be wasted. This is consistent with what it means to be a scientist.

Other research scientists use animals not as objects of study but as tools or sources of research material. The obligation of this scientist is the same obligation he or she has to the usage of other tools and materials, with the understanding that animals constitute a very special class of tools. The proper use of animal "tools" implies the obligation to provide tolerable noise levels, housing congenial to the species, and opportunities for play and social interaction for individual animals. But at least, preventing the death, suffering, and distress of research animals is no less important than preventing the loss of other important tools and materials.

CONCLUSION

It is not necessary to give nonhuman animals moral protections equal to the moral protections enjoyed by those within the moral community to show that we need to justify our use of them. Research scientists have moral obligations to justify, minimize, and prevent the death, suffering, and distress of research animals. The way this plays out in the typical research setting includes some of the following:

1. The role of the Institutional Animal Care and Use Committee (required by federal law at any U.S. research site using animals) includes asking questions of investigators that go beyond current federal regulations. The IACUC is responsible for monitoring how investigators might justify, minimize, or prevent death, suffering, and distress to research animals.
2. Research scientists have a positive obligation to seek alternatives to animal use (thus preventing and minimizing death, suffering, and distress).
3. This analysis of research animal use also provides criteria by which to judge the legitimacy of nonresearch use of animals:
   a. How justified is the use when considered by knowledgeable members of the moral community? Although there may never be full agreement as to what constitutes justified use, the use will be more or less justified, based on the potential to benefit human life.
   b. Who is the moral agent causing animal death, suffering or distress? What are this person's role-related responsibilities to

minimize and prevent the death, suffering and distress of animals in his or her care?

The implication is that the person who is in a special position to cause death, suffering, and distress to animals has a concurrent obligation for justifying, minimizing, and preventing these evils for the animals in their care. This should hold whether we are examining the role of research scientists, or those who kill for food, or of pet owners.

Notes

1. T. Machan, "Do animals have rights?" *Public Affairs Quarterly* 5(2) (1991):168.

2. S. Sapontzis, *Morals, Reason, and Animals* (Philadelphia: Temple University Press, 1987), 216.

3. P. Singer, "All animals are equal." in *Animal Rights and Human Obligations* (Englewood Cliffs, N.J.: Prentice Hall, 1976), 154.

4. B. Gert, *Morality: A New Justification for the Moral Rules* (New York: Oxford University Press, 1988), 85.

5. C. Stone, "Should trees have standing." *Sanctuary* (Feb./March 1988):15.

6. See, for example, the case reported in the lay press in March 1992 of Theresa Ann Pearson, an anencephalic, who, although born alive, was considered for donation of life-sustaining organs.

7. For example, certain expensive medical treatments (such as organ transplants) in the United States are routinely withheld from patients who meet clinical requirements but who lack the necessary financial resources. As a society, we may agree that it's not "right" that insurance or the ability to attract media attention makes the difference between people who live and people who die of identical maladies. Nevertheless, we accept the immoral but economic reality. Our willingness to tolerate the present system does not imply an actual or perceived acknowledgment of morality.

8. Gert, *Morality,* 85.

9. Gert, *Morality,* 85.

10. P. Harrison, "Do animals feel pain," *Philosophy* 66(1991):38.

11. See, for example, F. Keefe, R. Fillingim, and D. Williams, "Behavioral assessment of pain: Nonverbal measures in animals and humans," *ILAR News* 33(1–2)(1991):3–13.

12. B. Rollin, *The Experimental Animal in Biomedical Research, Vol. I, A Survey of Scientific and Ethical Issues for Investigators* Boca Raton, Fla.: CRC Press, 1990), 27.

13. P. Osgood, "The assessment of pain in the burned child and associated studies in the laboratory rat," *ILAR News* 33(2)(1991).

14. Sapontzis, *Morals, Reasons, and Animals,* 221.

15. Conduct disorder if diagnosed before the age of 15; antisocial personality disorder if diagnosed later, according to the *Diagnostic and Statistical Manual* of the American Psychiatric Association.

16. Rollin, *The Experimental Animal,* 27.

17. Committee on Science, Engineering, and Public Policy, *On Being a Scientist: Responsible Conduct in Research,* second edition (Washington, D.C.: National Academy Press, 1995), 1.

CASES FOR CONSIDERATION

CASE 1[1]

Among the risks faced by patients receiving artificial implants are infection and rejection of the foreign material. Infection is a particular problem with orthopedic fixation pins, since they often extend outside of the body's surface. Bacterial infection can travel along the fixation device, and can become very difficult to treat with conventional antibiotic therapy.

The Department of Orthopedics was studying the development of an antibacterial coating on orthopedic fixation pins. They developed a research plan to attempt to show that subcutaneous implanted stainless steel pins could serve as a model for evaluating the efficacy of different pin coatings. As a first step, the team could inoculate bacterial strains isolated from humans with orthopedic pin infections alongside an implanted uncoated pin. If the model proved workable, and the bacterial recovery procedures proved to be adequate, studies could proceed with coated pins.

The IACUC protocol requested ten rabbits. Using two of the rabbits, the research team would subcutaneously implant a 1-inch piece of a standard stainless-steel ⅛-inch orthopedic pin in each animal's flank. The team would first anesthetize the rabbits with a ketamine-xylazine combination given intramuscularly, give the skin a surgical prep, and insert the pin through a small stab incision. One week after implantation, and only if there was no overt infection, they would infect 1×10^6 bacteria in 0.1 ml of saline around the incision site. After another week, they planned to reanesthetize the rabbit, disinfect the site, reopen the incision, and take and culture a sterile swab of the implant area. Finally, they would remove the implant and culture it two weeks later, at the time of euthanasia.

If during the course of experimentation a rabbit lost more than 20% of its initial weight (based on twice weekly weighings) the team planned to remove the pin and give the rabbit proper veterinary care. If the study was successful, as determined by the recovery of the injected bacteria from the pin, then the research would progress using four different pin coatings on groups of two animals.

Questions for Discussion

1. Can this study be done using in vitro techniques?
2. Will there be a host reaction to the materials?
3. What is the potential for pain and how will it be addressed?

4. Twenty percent loss of body weight is excessive. What other, less distressful, criteria could be used to remove an animal from study?
5. Are the numbers of animals appropriate? Either too many or too few?

CASE 2[2]

A major grant deadline was fast approaching so the IACUC was reviewing a plethora of protocols. The process was going smoothly until they reviewed Dr. Smith's protocol. Dr. Smith intended to anesthetize rats for surgery using 40 mg/kg of ketamine with five mg/kg of xylazine/kg body weight, intramuscularly. The surgical procedure involved making a scalp incision, moving the cranial muscles to expose the skull, and placing electrodes in the brain. Smith planned to anchor the electrode assembly to the skull with dental acrylic—standard procedure, he stated—then close the wound with surgical clips, dust it with an antibiotic powder, and allow the animal to recover. No further treatment was necessary, Smith wrote, because the brain has no pain sensation of its own.

Questions for Discussion

1. How good is the protocol description of the surgical procedure? Is the description of the absence of pain receptors in the brain adequate, or is there potential for pain in other aspects of the procedure?
2. What criteria will be used to assure that the animal is not in pain and/or determine the necessity for the use of analgesics?
3. What is being done to assure aseptic technique? What type of post-op monitoring is the investigator using?

CASE 3[3]

Tempers were short at the Great Eastern University IACUC monthly meeting. There were only five protocols to review, but the committee had discussed Stan Handleman's allograft study for half an hour, and no end was in sight.

Dr. Handleman, a noted surgeon, had an approved protocol to perform canine pancreatic allografts, which involved implanted pancreatic tissue from a purpose-bred donor dog under the capsule of one kidney while using the other kidney as a control. He would remove the recipient dog's pancreas, euthanize the donor dog, which provides the pancreatic tissue for the allograft, and donate the donor dog's tissue to unrelated research projects.

Immediately after surgery, the recipient would receive slightly different experimental drugs, with and without immunotherapy, to aid the survival of the allograft. During the postoperative period, Handleman would monitor the recipient's blood glucose, general health, and various enzymes. He would euthanize the recipients at specific times, or euthanize them earlier if they were not faring well, and remove the kidneys for histologic and other examinations. The maximum allowable survival time was 30 days with a combination of drug and immunotherapy. Using that combination, dogs did well.

The success of the study prompted Handleman to request a protocol modification, extending the study to 60 days. Instead of implanting pancreatic tissue in one site, he would implant it in two separate areas of the same kidney. Thirty days after the first surgery, he would perform a partial nephrectomy, removing tissue from the implant site. Thirty days later, he would euthanize the dog and remove both kidneys. This would allow him to do a time-course study in the same animal.

Questions for Discussion

1. Is it better to minimize the pain that each dog experiences by increasing the number of animals or is it better to minimize the number of animals used, even if it means the animals used have a greater potential to experience pain?
2. Who should assess the amount of pain each animal is likely to experience? How might the issues of potential for pain, recognition of pain, and treatment of pain be discussed in this protocol?
3. What is the purpose of the study and what benefits are likely to accrue from this study?

Notes

1. J. Silverman, "Pinning down a protocol," *Lab Animal* 22(10) (1993):23.
2. J. Silverman, "Protocol review, Brain pain," *Lab Animal* 23(1) (1994):22.
3. J. Silverman, "Protocol review, Allograft stalemate," *Lab Animal* 23(6) (1994): 22.

Regulations and Requirements

B. TAYLOR BENNETT

INTRODUCTION

Since the ultimate responsibility for compliance with regulations that affect the care and use of animals lies with the investigator, it is important that he/she have a working knowledge of the basic regulatory requirements. In this manual, the types of regulations will be discussed under two broad general headings:

1. Involuntary
2. Voluntary

Involuntary regulations can be defined as those required by law or set forth as a condition of funding. There are four types of regulatory controls which can be considered as involuntary:

1. The Animal Welfare Act (AWA)
2. The Public Health Service policy
3. The Good Laboratory Practices Act
4. The requirements of private funding agencies

Voluntary regulations can be defined as those that an individual or institution adheres to as part of their overall commitment to research and academic excellence. There are two types of regulatory controls which can be considered as voluntary:

Reprinted from B. T. Bennett, M. J. Brown, and J. C. Schofield, eds., *Essentials for Animal Research: A Primer for Research Personnel,* 1–7, by permission of the National Agriculture Library, Beltsville, Md. Copyright 1994.

1. Accreditation by the American Association for the Accreditation of Laboratory Animal Care (AAALAC)
2. Requirements of individual users

INVOLUNTARY REGULATIONS

Animal Welfare Act

The Animal Welfare Act was first passed August 24, 1966, as PL-89-544. It was entitled the "Laboratory Animal Welfare Act" and authorized "The Secretary of Agriculture to promulgate such rules and regulations, and orders as he may deem necessary to effectuate the purposes of this Act." The purposes of the original act were to:

1. Protect the owners of dogs and cats from theft of such pets.
2. Prevent the sale or use of dogs and cats which had been stolen.
3. Insure that certain animals intended for use in research facilities were provided humane care and treatment.

In charging the Secretary, Congress specifically prohibited the promulgation of rules, regulations, or orders which would interfere with the conduct of actual research. Determination of what constituted actual research was left to the discretion of the research facility.

The original Act covered non-human primates, guinea pigs, hamsters, rabbits, dogs, and cats. Humane treatment was required while they were at the dealers or research facility and while being transported by dealers. Dealers were required to be licensed. Research facilities which used, or intended to use, dogs or cats and either purchased them in commerce or received any federal funds were required to be registered.

The Secretary also established regulations and standards for the implementation of unannounced facility inspections and for the maintenance of specific records by dealers and research institutions. Responsibility for administering the Act was delegated within the United States Department of Agriculture (USDA) to the Administrator of the Animal and Plant Health Inspection Service (APHIS). Enforcement duties are the responsibility of the APHIS Deputy Administrator for Regulatory Enforcement and Animal Care (REAC). The actual inspections are conducted by 42 Veterinary Medical Officers working under one of the five REAC Sector Supervisors. The Sector offices are located in Fort Worth, Texas, Tampa, Florida, Annapolis, Maryland, Minneapolis, Minnesota, and Sacramento, California.

In 1970 the original Act was amended (PL- 91-579) and renamed the Animal Welfare Act. The amended Act covered broader classes of animals and included those used in exhibitions and sold at auction and regulated anyone involved in these activities. The definition of an animal was expanded to include all warmblooded animals. The definition of a research facility was expanded to include those institutions using covered live animals and not just dogs and cats. These facilities were required to file an annual report. Civil penalties were also added for refusing to obey a valid cease and desist order from the Secretary. The term "handling" was added to the basic categories for which standards were to be created and the phrase "adequate veterinary care" was broadened to include the appropriate use of anesthetics, analgesics, and tranquilizers.

The intent of the original Act to prohibit interference with research was clarified and the Secretary was enjoined from directly or indirectly interfering with, or harassing in any manner, research facilities during the conduct of actual research or experimentation. The determination of when actual research was being done was still left to the discretion of the research facility itself.

In 1976, the Animal Welfare Act was further amended to enlarge and redefine the regulation of animals during transportation and to combat the use of animals for fighting. Essentially the Act was broadened to include all forms of commercial transportation of animals and required all carriers and intermediate handlers who were not required to be licensed under the Act to register with the USDA. It also expanded the definition of a dealer and extended the record keeping requirements to carriers and intermediate handlers.

In 1976, the Secretary also promulgated regulations which specifically excluded rats, mice, birds, horses, and farm animals from the definition of an animal. This exclusionary language effectively excludes over 80 percent of the animals currently used in research, teaching, and testing from coverage under the Animal Welfare Act.

In 1985 the Act was further amended with the passage of the Food Security Act of 1985 (PL-99-198), which contained an amendment entitled the "Improved Standards for Laboratory Animals Act." This amendment when fully implemented will strengthen the standards for providing laboratory animal care, increase enforcement of the Act, provide for collection and dissemination of information to reduce unintended duplication of experiments using animals, and mandate training for those who handle animals.

The most recent amendment to the AWA also includes development of standards: for the "exercise of dogs," for "provision of a physical environment which promotes the psychological well-being of primates," for limitation of multiple survival surgeries, and to require the investigator to consult with a veterinarian in the design of experiments which have the potential for causing pain to insure the proper use of anesthetics, analgesics, and tranquilizers. Each research facility will have to show upon inspection, and include in their annual report, assurances that professionally acceptable standards for the care, treatment, and use of animals are being used during the actual research or experimentation. As part of these standards, the investigator is required to consider alternative techniques to those which might cause pain or distress in the experimental animals.

The 1985 amendment requires the Chief Executive Officer of each research facility to appoint an Institutional Animal Committee consisting of at least three members including a doctor of veterinary medicine and one member who is not affiliated with the institution. The regulations promulgated to implement the amendment designate this committee as the Institutional Animal Care and Use Committee (IACUC) and charge it to act as an agent of the research facility in assuring compliance with the Act. The Committee is required to inspect all animal facilities and study areas at least once every six months, and to review the condition of the animals and the practices involving pain to the animals, to insure compliance with the regulations and standards promulgated under the Act. The Committee is also required to review once every six months the research facility's program to assure that the care and use of the animals conforms with the regulations and standards. The Committee must file a report of its inspection with the Institutional official of the research facility. If significant deficiencies or deviations are not corrected in accordance with the specific plan approved by the Committee, the USDA and any Federal funding agencies must be notified in writing.

The Committee must also review and approve all proposed activities involving the care and use of animals in research, testing, or teaching procedures and all subsequent significant changes of ongoing activities. As part of this review, the Committee must evaluate procedures which minimize discomfort, distress, and pain and [ensure] that when an activity is likely to cause pain a veterinarian has been consulted in planning for the administration of anesthetics, analgesics, and tranquilizers and that paralytic agents are not employed except in the anesthetized animal. The IACUC must also determine that animals which experience severe or chronic pain are euthanized consistent with the design of the study, that the living conditions meet the species needs, that necessary medical care will be provided, that all procedures will be performed by qualified individuals, that survival surgery will be performed aseptically, and that no animal will undergo more than one operative procedure that is not justified and approved. Methods of euthanasia must be consistent with the definition contained in the regulations.

The IACUC must also assure on behalf of the research facility that the principal investigator considered alternatives to painful procedures and that the work being proposed does not unnecessarily duplicate previous experiments. To provide this assurance the Committee must review the written narrative description provided by the investigator. This description must include the methods and sources used in determining that alternatives were not available.

In reviewing proposed activities and modifications, the IACUC can grant exceptions to the regulations and standards, if they have been justified in writing by the principal investigator.

In addition to the above requirements, the research facility is required to provide training in the following areas to scientists, animal technicians, and other personnel involved with animal care and treatment:

1. Humane practice of animal maintenance and experimentation.
2. Research or testing methods that minimize or eliminate the use of animals or limit pain or distress.
3. Utilization of the information service of the National Agricultural Library.
4. Methods whereby deficiencies in animal care and treatment should be reported.

The regulations require that each research facility establish a program of adequate veterinary care that includes: appropriate facilities, personnel, and equipment; methods to control, diagnose, and treat diseases; daily observation and provision of care; guidance to personnel on the use of anesthetic, analgesic and euthanasia procedures and pre- and post-procedural care. Specific requirements for maintaining records and filing annual reports are included in the regulations along with a miscellaneous section containing a variety of requirements to which a research facility must adhere.

Public Health Service Policy

The Public Health Service Policy on Humane Care and Use of Laboratory Animals can be found in Chapter 4206 of the NIH Manual and Chapter 1-43 of the PHS Manual. The NIH originally initiated the Policy in 1971. It was extended to all PHS activities January 1, 1979, and was revised in the spring of 1985 with implementation to be effective January 1, 1986. With the passage of the Health Research Extension Act of 1985 (PL-99-158), the Policy was further revised and the Director of the NIH was required by law to establish guidelines which heretofore had only been a matter of PHS policy. An additional revision was released in September 1986 which reflected the changes required by this Act.

Under the PHS policy, each institution using animals in PHS-sponsored projects must provide acceptable written assurance of its compliance with the Policy. In this Letter of Assurance the institutions must describe:

1. The institutional program for the care and use of animals.
2. The institutional status.
3. The Institutional Animal Care and Use Committee (IACUC).

The *institutional program* must include a list of every branch and major component, the lines of authority for administering the program, the qualifications, authority, and responsibility of the veterinarian(s), the membership of the Institutional Animal Care and Use Committee, and the procedures which they follow must be stated. The employee health program must be described for those who have frequent animal contact. A training or instruction program in the humane practices of animal care and use must be available to scientists, animal technicians, and other personnel involved in animal care, treatment, and use. The gross square footage, average daily census, and annual usage of each animal facility must be listed.

The *institutional status* must be stated as either Category one (1) (AAALAC accredited) or Category two (2) (nonaccredited). Institutions in Category two (2) must establish a reasonable plan with a specific timetable for correcting any departures from the recommendations in the *Guide for the Care and Use of Laboratory Animals* (86-23).

The *IACUC* must be appointed by the Chief Executive Officer and consist of at least five members; one of whom is a veterinarian with program responsibility, a practicing scientist, an individual whose expertise is in a non-biological science, and an individual who is not affiliated with the institution. This Committee must use the *Guide* to review the animal facilities and the institutional program for humane care and use of animals at least once every six months and prepare reports of these evaluations for the responsible institutional official. The Committee must review and approve animal-related components of proposals and significant modifications made in ongoing activities involving the care and use of animals. The Committee is responsible for reviewing concerns involving the care and use of animals and making recommendations to the institutional official regarding any aspect of the animal program, the facilities, or the personnel training. They are also authorized to suspend activity involving the care and use of animals as set forth in the PHS Policy.

In reviewing the animal care and use component of a proposal, the IACUC must confirm that the project will be conducted in accordance with the AWA and consistent with the recommendations in the *Guide.* In addition, all procedures are reviewed to assure that pain or distress will be minimized and that (when necessary) appropriate anesthetics, analgesics, and tranquilizers will be used The living conditions and medical care available must be appropriate for the species used, and personnel conducting the procedures must be appropriately trained and qualified. Methods of euthanasia should be consistent with the recommendations of the American Veterinary Medical Association Panel on Euthanasia.

The investigator is responsible for completing a proposal in accordance with recommendations in the PHS Policy and the instructions contained in the PHS 398 application packet. As of October 1988, the instructions for completing 398 can be found in two locations within the application package. On page 6 the research investigator's responsibilities for assuring the humane care and use of animals are clearly addressed. Detailed instructions for completing Section F of the Research Plan which describes the use of Vertebrate Animals can be found on page 21.

The institution is responsible for maintaining all the necessary records to document compliance with the PHS Policy and for filing annual reports which detail any changes in the program and indicate the dates of the semi-annual inspections and programmatic reviews.

The PHS Policy described above is intended to implement and supplement the "U.S. Government Principles for the Utilization and Care of Vertebrate Animals in Testing, Research and Training." The nine principles are published in the PHS Policy and in the Appendix of the

Guide. All those responsible for the design, supervision, and review of the animal care and use component of a proposal should be familiar with this document.

Good Laboratory Practices

In 1978 the Food and Drug Administration adopted the Good Laboratory Practices rules, which applied to all regulated parties who conduct nonclinical safety assessment studies. The rules require the creation of Standard Operating Procedures for all aspects of the study including animal care and use. A Quality Assurance Unit must be established to conduct internal inspection of practices and records to insure compliance with established policies and procedures. In general the recommendations contained in the *Guide* would suffice in terms of animal care when adherence is properly documented.

Private Funding Agencies

In recent years the requirements of many private funding agencies which fund research projects involving the care and use of laboratory animals have changed. It is important to obtain the requirements from the agency before spending time preparing a proposal. Some of these agencies not only require review of the proposal by the IACUC, but require proof of accreditation by AAALAC. In many instances, the proposals must be reviewed and approved prior to submission.

VOLUNTARY REGULATIONS

American Association for the Accreditation of Laboratory Animal Care (AAALAC)

AAALAC was originally chartered April 30, 1965, as a voluntary organization that accredited institutional programs of animal care and use. AAALAC is governed by a Board of Trustees composed of representatives of 32 professional organizations. An 18-member Board-appointed Council on Accreditation makes recommendations based on the results of site visits to evaluate an institution's compliance with the recommendations contained in the *Guide*. This is a peer review process in which standards are being continually upgraded to reflect current knowledge in laboratory animal medicine and science. In its accreditation program the AAALAC Council uses the *Guide* more as a compilation of regulatory "standards" and not as a set of "recommendations."

Since the AAALAC accreditation program and the *Guide* are so closely linked, a brief review of the *Guide*'s history and its current contents are warranted. In 1963 the first *Guide for Laboratory Animal Facilities and Care* was published by the Institute for Laboratory Animal Resources (ILAR) under a contract from NIH. Since its original release the *Guide* has been revised in 1965, 1968, 1972 (when the title was changed to the *Guide for the Care and Use of Laboratory Animals*), 1978, and 1985. In the most recent revision, the organization of the chapters was changed to reflect the increasing role and responsibility of the institutional program in establishing acceptable standards for the care and use of laboratory animals. The first chapter is now Institutional Policies. The remaining four chapters are Laboratory Animal Husbandry, Veterinary Care, Physical Plant, and Special Considerations. Prior to an AAALAC site visit, each institution is required to prepare a description of the institutional facilities and programs using the AAALAC Outline for Description of the Institutional Animal Care and Use Program, which follows the *Guide*'s chapter headings.

Once accredited, an institution must submit an annual report describing changes in the program and facilities and documenting the annual usage of animals. Site visits occur at least every three years, and these visits consist of an inspection and review of policies, procedures, and facilities which comprise the animal care and use programs inclusive of selected animal usage areas. Should deficiencies be identified in a previously accredited program, the institution is either granted a probationary period in which to make specified changes, or if the deficiencies are major, accreditation could be withdrawn.

Individual Users

The instructions for completing PHS 398 clearly define the roles and responsibilities of the investigator in assuring proper care and usage of laboratory animals. In addition to this requirement, it should be understood that any type of care or use of an animal which results in the creation of nonexperimental variables can potentially compromise the integrity of an entire project. As part of their commitment to scientific excellence, the users should provide the impetus for setting and maintaining high standards for the care and use of laboratory animals within their individual and collective institutions. Failure to do so invites increased internal and external regulatory requirements which can drain limited institutional research resources. Good animal care is good science; the practice of good science should be the primary goal of all who have chosen careers in the scientific community.

SUMMARY

In summary, the regulations that affect the use of animals in research, teaching, and testing programs are numerous. A working knowledge of the applicable regulations is necessary if the principal investigator is to ensure that proposals for funding contain the necessary information and to assure that the conduct of all research proposals is in compliance with the requirements of the regulatory and funding agencies. While the ultimate responsibility for compliance rests with the principal investigator, institutional policies should be designed to provide those responsible for compliance with the necessary resources to do so.

References

Application for Public Health Service Grant, PHS, 398. OMB No. 0925-001. Revised October, 1988.

Animal Welfare Act (Title 7 U.S.C. 2131-2156), as amended by PL-99-198. Beltsville, Md. (December 12). 1986.

Guide for the Care and Use of Laboratory Animals. NIH Publication No. 86-23.

Public Health Service Policy on Humane Care and Use of Laboratory Animals. Revised Office for the Protection of Research Risks, Bethesda, Md. (September). 1986.

"Non-clinical laboratory studies. Good laboratory practice regulations." *Federal Register.* December 22 (Part II):59986–60026. 1978.

Public Law 99-198. Code of Federal Regulations. Title 9, subchapter A, "Animal welfare." 1989.

Townes, J. "Federal regulations, an overview." *Lab Animal.* 9(4):16–22. 1980.

An Introduction to the Philosophical Presuppositions of the Animal Liberation/Rights Movement

RICHARD P. VANCE

The animal liberation/rights movement (ALRM) poses a significant threat to the future of medical research. This threat is far greater than currently recognized by most physicians.[1, 2] One of the strengths of the ALRM is its multifaceted nature.[1–7] Some activists attack research by picketing facilities, intimidate researchers via threatening mail or phone calls, or destroy property and threaten lives.[3–5, 8] Others attack research through political, regulatory, and bureaucratic avenues, or use intimidating legal maneuvers.[1, 2, 4, 8] Others criticize research by posing as objective professional organizations.[1, 2, 9] Finally, some professional philosophers provide justifications for ALRM activities.[10–20]

Defenders of research often miss the mark in their analysis and criticisms, because political terrorist activities tend to draw attention away from careful consideration of ALRM's philosophical presuppositions.[21–23] As a result, we have consistently misjudged and underestimated the potential danger of the ALRM. From the beginning, the ALRM has been a *liberation movement* in which philosophy and politics are explicitly connected. The philosophical foundations of the movement can be used to endorse a wide range of political acts, including violence, as long as they receive the right kind of response among the general public, or are undertaken for the right reasons.[20[xii–xiii]] If we are to understand why the ALRM acts as it does, and to respond to it effectively, we need to understand more adequately its basic philosophical foundations.

The two most important philosophical figures in the ALRM are Peter Singer and Tom Regan. Peter Singer, author of *Animal Liberation*,[20] is considered by Ingrid Newkirk and Alex Pacheco of People for the Ethical Treatment of Animals to be the "father" of the modern animal liberation movement. Regan's most influential book is *The Case for Animal Rights*.[13] There are, of course, other philosophers who support the ALRM,[10, 11, 15–18] but Singer and Regan provide the most widely recognized philosophical analyses to date. In many ways, their work set the stage for everything else that has followed in the ALRM.

Because defenders of the research have focused their attention on the activities and slogans of the ALRM, their efforts to rebut the movement have too often been guided by myths or oversimplifications. The three most persistent myths are that (1) the ALRM is motivated primarily by emotion, sentiment, or mysticism; (2) the ALRM denies all differences between humans and animals; and (3) the ALRM has a seamless, coherent philosophy (i.e., there are no serious conflicts at a fundamental philosophical level). This essay, by examining the philosophical foundations for the ALRM, exposes these myths and provides useful criticisms of the ALRM's philosophical presuppositions.

MYTHS ABOUT THE ALRM MOVEMENT

Myth 1: ALRM as Emotionalism

It often appears that the ALRM is motivated by nothing more than exaggerated emotional or mystical sentiment. Many believe ALRM members have never outgrown a Walt Disney image of animals. If only animal activists could understand rationally the dangerous world we live in, and the difficulties medical science faces in trying to alleviate suffering and cure disease, they would cease to be critical of animal use in research. Although this may be true of some ALRM supporters, ALRM philosophers don't provide such easy targets. Regan and Singer base their positions on modern secular reasons. Both eschew arguments based on religious presuppositions. Indeed, Singer explicitly denies any religious motivation,[20[ii], 24] and in *Animal Liberation* points out that he has never "been inordinately fond of dogs, cats, or horses in the way many people are."[20[ii]] He is "not especially 'interested in' animals" if that means having sentimental attachments to them.[20[ii]] For Singer, emotion follows reason. Only *after* a situation has been shown by rational analysis to be immoral should emotion enter (e.g., as justified outrage). Similarly, Tom Regan argues that animal activism derives from rational arguments: "The idea of animal rights has reason, not just emotion, on its side."[20[14]]

We ought to take Singer and Regan seriously about their claims to being rational because their methods are typical of analytical ethics. Analytical ethics is both a complex tradition and one of the most common types of philosophical ethics in the United States and Great Britain.[25–30] That part of the analytical tradition of concern in this essay has its most immediate roots in the 1940s. Analytical ethics was a response to arguments that ethics was merely the expression of emotion or rhetoric. Ethics is objective, analytical ethicists argued, because it is rooted in reason and in the very logic of our language. Analytical ethicists sought by logical or linguistic analysis to make sure that ethical terms and concepts are used consistently. Without analysis, ethical terms and concepts are too often applied in ambiguous or contradictory ways. Indeed, it was inconsistent or imprecise use that led to the false conclusion that ethics is not objective.

Logical consistency and universalizability are two hallmarks of analytical ethics. Universalizability simply means that any ethical prescription (e.g., you ought to tell the truth) is applicable to everyone in relevantly similar circumstances. If one makes a different moral judgment in two different cases, one must demonstrate a morally relevant difference between the two cases. If a duty can be extended to include more people or situations, then the logic of analytical ethics demands that it be extended. The broadest, and most consistent, use of a term or concept is the goal for ethics, because objectivity cannot be achieved otherwise.

Both Singer and Regan present mainstream examples of analytical ethics. Methodologically, neither is radical or bizarre. In *Animal Liberation,* Singer is a very typical utilitarian. One should act, Singer argues, to bring about the best balance of good and bad consequences for everyone affected. Furthermore, Singer is guided by the basic moral principle of equal consideration of interests.[20] Not to be hurt is an interest that is the same for all sentient creatures (i.e., organisms that are responsive to or conscious of sense impressions). Therefore, since animals can feel pain and like humans naturally seek to avoid it, all sentient beings ought to be considered as equals with respect to the infliction of pain. A difference of species is not a morally relevant distinction; the only relevant boundary is the limit of sentience.[20] To act purely on the basis of species difference with regard to the infliction of pain is, for Singer, no different than to act solely on the basis of race or sex.

Singer is heavily indebted to Jeremy Bentham, the nineteenth-century British philosopher and father of utilitarianism. Regarding the treatment of animals Bentham argued, "The question is not, Can they reason? nor Can they talk? but, Can they suffer?"[31] Based on utilitarian presuppositions, therefore, Singer argues,

> If some experiment would hurt a pig and a human to the same extent, and there were no other relevant consequences of the use of the human being or the pig, it would be wrong to say to use the pig because the suffering of the pig counts less than the suffering of the human being.[32[48]]

Generically, Singer's argument is no different than arguments we make about equal treatment of unequal human beings. Even though some humans are less intelligent or able than others, we would not consider the infliction of pain on such humans less important than the infliction of pain on more gifted people. To the extent that we all have equal interest in not being tortured, we are equal. Singer merely extends this "interest" to animals. Singer's argument is certainly contestable, as will be demonstrated below. Yet Singer's principle of equal consideration of interests is precisely what one would expect a consistent utilitarian philosopher to say about the universalizability of equal interests.

In contrast to Singer, Regan argues that animals have moral rights. He bases this claim on the concept of "inherent value."[12–14, 19] Inherent value places a burden on others not to interfere with or to disturb that value. Regan establishes this moral right as a basic moral intuition. Although he does not sufficiently acknowledge it in his animal-rights writings, Regan's thinking is heavily influenced by the British intuitionist G. E. Moore. Moore is a central figure in the tradition of philosophical analysis and is most typically identified as the philosopher who described the notion of intrinsic values: simple nonnatural, unanalyzable properties that are known clearly and immediately to us through our moral intuition.[27, 33, 34] Regan distinguishes his intuitionism from Moore's by arguing that intuitions are

> not our "gut responses," nor are they merely expressions of what we happen to believe; they are our *considered* beliefs, beliefs we hold when, and only when, we have done our best to be impartial, rational, cool, and so forth.[13[134]]

Despite this distinction, Regan's inherent value serves a function very similar to Moore's intrinsic value. Inherent value is most basically justified, Regan argues, because it "provides a theoretical basis for avoiding" all the problems that appear in other ethical positions.[13[247–48]]

How does Regan move from his defense of inherent value to animal rights? Once inherent value is postulated, Regan uses what he calls a "subject of a life" criterion to determine which organisms are to count as moral beings.[13[247–48]] A "subject of a life" is a "conscious creature having an individual welfare that has importance to it whatever its usefulness to others."[19[22]] Any such an organism has, Regan claims, inherent value[13[243–48]] and is due the respect accorded to moral beings that have moral rights.[13[279–80]] This means at the very least that any mammal after the age of infancy has "moral rights." From this complex intuitionist position, Regan argues that we ought not think about the consequences for human welfare when considering research using animals.[13[382–94]] Instead, our concern should be the inherent value of the animals. Inherent value (and the moral rights that attach to inherent value) should guide our moral action. Humans ought

> never to harm the individual merely on the grounds that this will or just might produce "the best" aggregate consequences. To do so is to violate the rights of the individual. This is why the harm done to animals in the pursuit of scientific purposes is wrong. The benefits derived are real enough; but some gains are ill-gotten and all gains are ill-gotten when secured unjustly.[32[49]]

Despite their differences, both Regan and Singer extend beyond the species barrier basic ethical principles, terms, and concepts. This extension is demanded by the logic of universalizability—to seek the broadest and consistent application of the principle that cases should be treated similarly unless there is a relevant difference between them. Despite their radical conclusions, both Singer and Regan are very much in the mainstream in their ethical methods. Consequently, the charge that they are motivated primarily by emotion is unconvincing to anyone who takes the time to read their work and is familiar with modern ethics.

Myth II: ALRM Denies the Differences Between Animals and Humans

The second myth is no less commonly asserted than the first. For many people, Ingrid Newkirk's statement that "[W]hen it comes to feelings, like pain, hunger, and thirst, a rat is a pig is a dog is a boy" seems to justify it.[35] Yet, Singer and Regan are careful to distinguish between animals and humans.[8] Even though animals have inherent value or deserve equal consideration of interests, animals are not equal in all respects to humans, even for moral evaluation. Singer, for example, says "it is worse to kill a normal adult human, with a capacity for self-awareness . . . than it is to kill a mouse."[20[19]] Singer says that "we may legitimately hold that there are some features of certain beings that make their lives more valuable than those of other beings. . . ."[20[19]] Normal adult humans have on balance more value than animals: "[N]ormally this will mean that if we have to choose between the life of a human being and the life of another animal we should choose to save the life of the human."[20[21]] Regan takes a similar position. Given a situation in which one must choose between the lives of human beings and an animal's life, "it is not speciesist to claim that the death of any one of these humans would be a *prima facie* greater harm. . . ."[13[351]] These conclusions are hardly counter-intuitive for most of us. But we typically believe that animal activists fail to make these basic distinctions. Yet, both Regan and Singer describe differences between humans and animals and argue that these differences make a difference in certain kinds of situations.

What sorts of situations are these? Most typically, only situations in which the life of a human being is in direct and immediate conflict with the life of an animal. Understanding the roles these conflicts play in Singer and Regan's philosophies is crucial. These conflicts uncover basic internal weaknesses in both positions. Even more, these same conflicts demonstrate how much tension there is between Regan and Singer. Unfortunately, defenders of medical research have heretofore left these issues unaddressed.

Myth III: ALRM Has a Monolithic Ideology

Despite recent claims by ALRM philosophers,[36, 37] there is no substantial convergence in ALRM philosophies, and nothing demonstrates the divergences better than the writings of Regan and Singer. It is no accident that Singer calls the movement he started the *animal liberation* movement, while Regan calls it the *animal rights* movement. Far from representing different facets of a monolithic ideology, Regan's and Singer's positions are not rationally reconcilable. This should not be surprising, since analytical ethicists are famous for providing devastating criticisms of rival analytical models.[38–44] Indeed, criticisms *against* a model are often far more convincing than any positive arguments *for* a model. Singer and Regan are no exception; Regan provides convincing criticisms of Singer, and Singer provides devastating criticisms of Regan. Each shows that his competitor's ethical model is inconsistent and leads to immoral or irrational outcomes. Singer and Regan, therefore, provide for us all the arguments needed to demonstrate the incoherence of their philosophical presuppositions.

For example, Regan shows that Singer's ethic is flawed by using a hypothetical story:

> My Aunt Bea is old, inactive, a cranky, sour person, though not physically ill. She prefers to go on living. She is also rather rich. I could make a fortune if I could get my hands on her money, money she intends to give me in any event, after she dies, but which she refuses to give me now. In order to avoid a huge tax bite, I plan to donate a handsome sum of my profits to a local children's hospital . . . Why not kill my Aunt Bea? Oh, of course I *might* get caught. But I'm no fool, and besides, her doctor can be counted on to cooperate . . .[19[20]]

Without some sort of categorical protection for the individual, Regan argues, utilitarianism might very well determine that it is morally correct to kill Aunt Bea. Indeed, argues Regan, utilitarianism provides no categorical protection against any kind of action, as long as the greater good for the greater number occurs as a result. Utilitarianism is flawed because it sanctions terribly immoral deeds in the name of the greater good.

In his own defense, Singer argues that utilitarianism is more flexible and reasonable than other alternatives, especially when dealing with animal experimentation.[32] For example, Singer argues that he *might* approve an experiment on one animal causing painless death, if the experiment would provide the knowledge to cure all forms of cancer, and if any human of a mental capacity similar to the animal could also be used[32[49]] Singer seems to believe this makes utilitarianism a *moderate* ALRM position, occupying a *reasonable middle ground.*[20, 32, 45] Regan responds that even this experiment would be immoral, because it would violate the basic notion of equal inherent value.[46] We would never allow humans to be used in this way, Regan says. Therefore, the rights of adult mammals must logically be protected in the same way.

Singer also provides strong criticisms of Regan's position. He does this by focusing on a different kind of case—a lifeboat case.[32, 46] In his writings,[13] Regan describes a situation in which four ordinary humans and a dog find themselves on an overcrowded lifeboat that is threatening to sink. Someone must be thrown overboard. Which one should it be? Regan argues that even though the dog and the humans each have the same inherent value, and an equal right not to be harmed, the magnitude of harm "is a function of the number and variety of opportunities for satisfaction it forecloses for a given individual."[13[351]] The harm of death for a normal adult human is greater than for an animal, Regan argues, because a human death takes away greater possibilities for satisfaction. Hence, Regan states: "No reasonable person would deny that the death of any of the four humans would be a greater *prima facie* loss."[13[324]]

More surprising still, Regan argues that even the presence of a million dogs would not make it right to throw one of the humans overboard.[13[325]] As long as the "intrinsic value" of each dog is considered one at a time—and each dog is thrown overboard one at a time—Regan argues that one may proceed until all 1 million dogs are thrown overboard. On this lifeboat, it is even permitted, Regan argues, to eat a dog to avoid starvation.[13[351–52]] The otherwise strict obligation to be a vegetarian is overridden in such a situation. To refuse to eat the dog would be to go beyond one's obligations. Not surprisingly, Singer cannot help but think that Regan's arguments display "speciesist" tendencies.[32[50]] Where would one stop, Singer wonders, if such killing can take place?

Regan responds that he still provides significant barriers against animal use in research. For example, suppose the same lifeboat circumstances, except that the humans have a degenerative brain disorder, while the dog is healthy.[13[385–94]] There is also in the boat an untried medicine that promises a complete cure for the disease, but might also cause death. To whom should it be given? Regan argues that it cannot be given to the dog. To do so would make the dog merely the means to a human end.

Why isn't medical research aimed at curing deadly diseases analogous to the first lifeboat example? The answer, Regan says, is that research *coercively* puts the animal at risk, "when the animal would not otherwise run this risk."[46] Medical research forces an organism with "intrinsic value" to be used merely for the ends of others.[46]

> Not even a single rat is to be treated as if that animal's value were reducible to his *possible utility* relative to the interests of others, which is what we would be doing if we intentionally harmed the rat on the grounds that this *just might* "prove" something, *just might* "yield" a "new insight," *just might* produce "benefits" for others.[13[384]]

Isn't that exactly what happens when the dog is thrown overboard to save human life? Regan responds "No!" because the dog would (presumably) suffer a similar risk by remaining on a sinking ship and because (presumably) the dog got on the lifeboat voluntarily.[46[56–57]] Hence, for example, if someone had coerced the dog onto the lifeboat, it would be immoral to throw the dog overboard.

For Singer, arbitrariness is inevitable for a moral-rights intuitionist. Like many utilitarians, Singer echoes Jeremy Bentham's claim that there are no rights other than those explicitly defined in law. Rights are

> the fruits of the law, and the law alone. There are no rights without law—no rights contrary to the law . . . no natural rights—no rights of man, anterior or superior to those created by the laws. The assertion of such right, absurd in logic, is pernicious morals.[47]

Bentham called moral rights—such as those defended by Regan—"nonsense on stilts."[48]

From these arguments, one can easily understand why the notion of a coherent ALRM philosophy is a myth. Singer and Regan show better than anyone else that there is no foundation to be found. Regan states that "it is difficult to exaggerate the radical moral difference between Singer's utilitarianism and the (my) rights view."[46[56]] Singer is no less critical: Regan's view "actually permits much more—in fact, literally infinitely more—animal experimentation than the utilitarian view."[46[57]] Yet, Regan and Singer have agreed that theoretical differences shouldn't prevent them from seeking a common practical goal: to attack the use of animals in research.[46[57]] The problem, of course, is that without a consensus on a coherent theoretical foundation, they have no moral justification for their common political aims. The value of their mutually devastating attacks is that there currently is no coherent philosophical argument that can provide a consensus for the ALRM.

WEAKNESSES SUMMARIZED

The failures of Regan and Singer are not peculiar to them, or caused by their lack of insight, ingenuity, or intelligence. Both are exceptionally good philosophers in the analytical tradition. They provide sophisticated defenses of their positions. Their problem is not emotionalism, irrationality, or ideological closemindedness. Their problem is in the limited nature of the philosophical tools they use. Their ultimate theoretical weaknesses are extremely common among analytical ethicists.[38–42]

Critics of analytical ethics like to point out that utilitarians never provide enough defense against terrible crimes committed in the name of the greatest good for the greatest number. There is, moreover, no theory-neutral, objective, or publicly agreed-on notion of utility. What counts as important to one person may not count the same for another. Furthermore, each person can also value things differently in different circumstances (e.g., age, occupation, intelligence, and the like). Ultimately, utilitarianism depends more on the person doing the valuing than on the value itself. There can be no other explanation for the vast differences among utilitarians. Some utilitarians like R. G. Frey, for example, *defend* medical research, and use exactly the same ethical model as Peter Singer.[12, 14, 49–52] Frey and Singer obviously *weigh* values quite differently.

Similarly, intuitionists like Regan cannot help but appear arbitrary. Intuitionists have a very difficult time explaining why so many otherwise ethically sophisticated and sensitive people disagree with one another. Regan intuits that animals have inherent value and extends moral rights to them. Carl Cohen argues that the imputation of moral rights depends on the capacity for free moral judgment and that animals therefore cannot have rights.[53] Since intuitionism supposes that all rational humans have the same capacity to intuit ethically appropriate values and obligations, moral disagreements, such as those between Regan and Cohen, can be explained by Regan in only two ways. First, Cohen (and those who would agree with him) has insufficiently exercised his intuitive capacity and is therefore irrational, and/or second, Cohen does not *want* to exercise his intuition and is therefore immoral. This explains, in part, the turbulent atmosphere surrounding all intuitionist moral arguments pitched primarily at the level of "rights" language. Singer and Regan display in their dispute many of the problems besetting analytical ethics. Unlike more substantive ethical traditions (e.g., religious or ethnic traditions), analytical ethics cannot draw on a rich array of sources—canonical texts, authoritative readings, overlapping (even contradictory) platitudes, interpretative communities, and the like.[54–56] In comparison with such traditions, analytical ethics is abstract and thin. Despite claims of rational consistency, no analytical ethical model has been able to achieve enough coherence to claim adequacy. At the same time, rival models provide such devastating criticisms that no model remains powerful enough to deserve allegiance.

CONCLUSIONS

The ALRM clearly focuses inordinate attention on biomedical research.[6] Why is this so? Singer tells us the real reason:

> American animal researchers are a smaller and politically less powerful group than American farmers, and they are based in regions where animal liberationists live. They therefore made a more accessible, and slightly less formidable opponent . . .[24[37]]

It is up to medical researchers, physicians, and those who value medical research to change Singer's assessment regarding our political power. We must respond in a forthright and honest way about what we do. We must continue to ensure that animal use in research is humane and appropriate. We must also quit deceiving ourselves about the nature of the ALRM. We must become familiar with the analytical structure of their philosophical presuppositions, if we are to expose their fundamental weaknesses.

We can no longer afford comfortable but erroneous myths that make the ALRM appear emotional, ignorant, and brainwashed. At its best, the ALRM is intelligent, politically astute, and resourceful. Nevertheless, there are deep philosophical divisions within the ALRM that vitiate their pretense of a unified movement. Recent philosophical efforts in the ALRM already seek ways to begin to overcome the weaknesses in Regan and Singer.[18, 36, 37] More careful attention needs to be paid to these efforts. The future of medical research could be in jeopardy if we fail to understand our most dangerous and sophisticated critics.

Notes

1. West, A., and H. A. Pincus. "Physicians and the animal rights movement." *New England Journal of Medicine.* 324:1640–43. 1991.
2. Smith, S. J., and W. R. Hendee. "Animals in research." *Journal of the American Medical Association.* 259:2007–8. 1988.
3. McCabe, K. "Who will live, who will die?" *Washingtonian.* 21:112–57. 1986.
4. Horton, L. "The enduring animal issue." *Journal of the National Cancer Institute.* 81:736–43. 1989.
5. McCabe, K. "Beyond cruelty." *Washingtonian.* 25:72–195. 1990.
6. Nicoll, C. S., and S. M. Russell. "Analysis of animal rights literature reveals the underlying motives of the movement: Ammunition for counteroffensive by scientists." *Endocrinology.* 127:985–89. 1990.

7. Dresser, R. "Animal experimentation." In *Biolaw.* 253–71. J. Childress, P. King, K. Rothenberg, W. Waddington, and R. Goare, eds. University Publications of America. Washington, D.C. 1986.
8. Horton, L. "A look at the politics of research with animals: Regaining lost perspective." *Physiologist.* 31(3):41–44. 1988.
9. Physicians Committee for Responsible Medicine, Resolution 109. In *Proceedings of the House of Delegates.* 392. June 24–28, 1990. American Medical Association. Chicago. 1990.
10. Clark, S. R. *The Moral Status of Animals.* Oxford University Press. New York. 1977.
11. Rachels, J. *Created from Animals: The Moral Implications of Darwinism.* Oxford University Press. New York. 1990.
12. Regan, T. *All That Dwell Therein: Animal Rights and Environmental Ethics.* University of California Press. Berkeley. 1982.
13. Regan, T. *The Case for Animal Rights.* University of California Press. Berkeley. 1983.
14. Regan, T., and P. Singer. *Animal Rights and Human Obligations.* Prentice Hall. Englewood Cliffs, N.J. 1989.
15. Rollin, B. E., and M. L. Kesel. *The Experimental Animal in Biomedical Research, I: A Survey of Scientific and Ethical Issues for Investigators.* CRC Press. Boca Raton, Fla. 1990.
16. Rollin, B. E. *The Unheeded Cry.* Oxford University Press. New York. 1989.
17. Rollin, B. E. *Animal Rights and Human Morality.* Prometheus. Buffalo, N.Y. 1981.
18. Sapontizis, S. F. *Morals, Reason, and Animals.* Temple University Press. Philadelphia. 1987.
19. Singer, P. *In Defense of Animals.* Harper & Row. New York. 1985.
20. Singer, P. *Animal Liberation.* Second edition. Random House. New York. 1990.
21. Newkirk, I. *Save the Animals.* Warner Books. New York. 1990.
22. *The Animal Rights Handbook.* Living Planet Press. Los Angeles. 1990.
23. Morgan, R. *Love and Anger: An Organizing Handbook.* Animal Rights Network. Westport, Conn. 1980.
24. Singer, P. "Unkind to animals." *New York Review of Books.* (February 2):36–38. 1989.
25. Wittgenstein, L. *Philosophical Investigations.* Third edition. Macmillan. New York. 1953.
26. Rorty, R. *The Linguistic Turn.* University of Chicago Press. Chicago. 1967,
27. Urmson, J. O. *Philosophical Analysis.* Oxford University Press. New York. 1956.
28. Anscombe, G. E. M. *Ethics, Religion and Politics.* University of Minnesota Press. Minneapolis. 1981.
29. Stroh, G. W. *American Ethical Thought.* Nelson-Hall. Chicago. 1979.

30. Goodpaster, K. E., ed. *Perspectives on Morality.* University of Notre Dame Press. Notre Dame, Ind. 1976.
31. Bentham, J. "Introduction to the principles of morals and legislation." In *The Works of Jeremy Bentham, I.* J. Bowring, ed. Russell and Russell. New York. 1962.
32. Singer, P. "Ten years after animal liberation." *New York Review of Books.* (January 17):46–52. 1985.
33. Moore, G. E., *Principia Ethica.* Cambridge University Press. New York. 1903.
34. Regan, T. *G. E. Moore: The Elements of Ethics.* Temple University Press. Philadelphia. 1991.
35. Miller, B. L. "Physicians and the animal rights movement." *New England Journal of Medicine.* 325:1581. 1991.
36. DeGrazia, D. "The moral status of animals and their use in research: A philosophical review." *Kennedy Institute of Ethics Journal.* 1:48–70. 1991.
37. Sumner, L. W. "Animal welfare and animal rights." *Journal of Medicine and Philosophy.* 13:159–75. 1988.
38. MacIntyre, A. *After Virtue.* Second edition. University of Notre Dame Press. Notre Dame, Ind. 1984.
39. MacIntyre, A. "Ought." In *Against the Self-images of the Age.* 136–56. University of Notre Dame Press. Notre Dame, Ind. 1978.
40. MacIntyre, A. "Does applied ethics rest on a mistake?" *Monist.* 67:498–513. 1984.
41. MacIntyre, A. "A crisis in moral philosophy: Why is the search for the foundations of ethics so frustrating?" In 18–35. H. T. Engelhardt and D. Callahan, eds. *Knowing and Valuing.* Hastings Center. Hastings-on-Hudson, N.Y. 1980.
42. MacIntyre, A. "How virtues become vices: Values, medicine and social context. In *Evaluation and Explanation in the Biomedical Sciences.* 97–111. H. T. Engelhardt, and S. F. Spicker, eds. Reidel. Dordrecht, The Netherlands. 1975.
43. Rorty, R. *Philosophy and the Mirror of Nature.* Princeton University Press. Princeton, N.J. 1979.
44. Williams, B. *Ethics and the Limits of Philosophy.* Harvard University Press. Cambridge, Mass. 1985.
45. Singer, P. "Utilitarianism and vegetarianism." *Philosophy and Public Affairs.* 9:325–37. 1980.
46. Regan, T., and P. Singer. "The dog in the lifeboat: An exchange." *New York Review of Books.* (April 25):56–57. 1985.
47. Bentham, J. "Pennominal fragments." In *The Works of Jeremy Bentham.* Vol. 3. 221. J. Bowring, ed. W. Tait. Edinburgh, Scotland. 1843–1859.
48. Bentham, J. "Anarchical fallacies." In *The Works of Jeremy Bentham.* Vol. 2. 501. J. Bowring, ed. W Tait. Edinburgh, Scotland. 1843–1859.
49. Frey, R. G. *Rights, Killing, and Suffering: Moral Vegetarianism and Applied Ethics.* Blackwell. New York. 1983.

50. Frey, R. G. *Interests and Rights: The Case Against Animals.* Oxford University Press. New York. 1980.
51. Frey, R. G. "Moral standing, the value of lives, and speciesism." *Between the Species.* 4:191–201. 1988.
52. Frey, R. G. "Animal parts, human wholes: On the use of animals as a source of organs for human transplantation." In *Biomedical Ethics Reviews, 1987.* 89–107. J. M. Humbert and R. A. Almeder, eds. Humana Press. Clifton, N.J. 1987.
53. Cohen, C. "The case for the use of animals in biomedical research." *New England Journal of Medicine.* 315:865–70. 1986.
54. MacIntyre, A. *Whose Justice? Which Rationality?* University of Notre Dame Press. Notre Dame, Ind. 1988.
55. MacIntyre, A. *Three Rival Versions of Moral Enquiry.* University of Notre Dame Press. Notre Dame Ind. 1990.
56. Hauerwas, S. *A Community of Character.* University of Notre Dame Press. Notre Dame, Ind. 1981.

9

Human Experimentation

JUDY E. STERN and KAREN LOMAX

I. THE FOLLOWING IS A HYPOTHETICAL CASE

Professor Harry Jacobson spent a long and successful career doing research on plant biochemistry. In the later years of his career he saw the culmination of a discovery he had made many years earlier. Professor Jacobson had isolated a plant substance that was subsequently found to inhibit proliferation of the immune cells of mice. After years of investigation the substance began to be considered for use in human subjects. Professor Jacobson, coming from a background in basic botanical research, had never considered the issues related to research on human subjects. He therefore left all aspects of the protocols on human experimentation to his medical collaborators. After several months of clinical trails the project was cited for the investigators' failure to obtain adequate informed consent from the human subjects enrolled in the study. Jacobson was shocked when the investigating committee cited him for misconduct along with his medical co-investigators.

Researchers working in psychology or biomedical science are motivated to learn about research involving human subjects because they are directly involved in such research. Researchers working in fields unaccustomed to human experimentation may be less inclined toward learning about such issues. Although researchers in some fields may not anticipate becoming involved in human subjects' research, interactions in today's science are too complex for lack of involvement to be a certainty. Regardless of one's anticipated research direction, an understanding of the fundamental obligations of researchers in relation to research on human subjects is important.

Much has been written on research with human subjects that is of particular interest to, and use for, clinicians. We review these issues;

however, our focus is more specifically on the responsibilities of the basic scientist who is involved with research on human subjects.

II. WHAT IS HUMAN EXPERIMENTATION?

Human experimentation is any manipulation, test, or procedure of an experimental nature performed on a human being as part of a scientific or social science investigation. Clinical protocols to test new medicines or procedures are a common form of human experimentation and have led to most of the major advances in medicine. Many of these involve medical protocols and involve medically trained personnel. Clinical studies such as these are often linked directly to clinical care. There are a number of issues to consider with regard to human experimentation, such as the adequacy of informed consent, patient risk versus benefit, pain and discomfort involved in different treatment options, randomization of patients into control and test groups, and justification for use of placebo controls.

Clinical psychiatry and clinical psychology experiments are also human experimentation. Another common form of human experimentation is found in psychology or sociology experiments that are performed outside of a clinical setting. In these experiments subjects may be exposed to questionnaires, cognitive tasks, or emotional tests. In contrast to medical protocols, these protocols often do not have the potential for immediate benefit to the subjects under study. While some of these experiments may appear to be relatively risk free for subjects, in fact potential for harm exists. The controversial experiments of Stanley Milgram are an example.[1] In these experiments two researchers participated in deceiving a study subject. One researcher pretended to receive an electric shock delivered by the study subject, who believed that he or she was participating in a conditioned learning experiment. No shock was actually given. Another researcher talked with the study subjects, encouraging them to continue giving shocks even when the subjects wanted to quit. Many of the subjects of these experiments suffered psychological harm as a result of discovering that they would willingly cause another human being harm in these circumstances. The experiments brought up many ethical issues involving the informed consent given by study subjects and the problems with using deception in a study design. They also tested the limits of how far the increased knowledge acquired in a research study can outweigh the risk to the study subjects.

Physiology experiments comparing responses to stress and exercise tests are also human experimentation. Many of these experiments involve the participation of medical personnel and may require the availability of back-up medical services.

All uses of human subjects in experiments are human experimentation, even though these uses may seem simple and undemanding. Drawing blood from volunteers constitutes human experimentation even when the volunteers are friends or colleagues. All such experimentation requires committee approval and the informed consent of subjects.

III. STANDARDS FOR RESEARCH ON HUMAN SUBJECTS

Standards for clinical and basic science researchers regarding the ethical design and conduct of research on human subjects are set out in several international codes and in U.S. federal statutes and regulations. All researchers should be familiar with these documents. The first of the international guidelines, the Nuremberg Code,[2] was drafted in 1949 shortly after the end of World War II. It is primarily a human rights document and was drafted by the judges presiding over the Nazi Doctors trial in response to the atrocities committed by Nazi physicians during unconscionable human experiments on inmates of concentration camps. In 1969, the World Medical Association drafted and adopted the Declaration of Helsinki[3] (most recently revised in 1989) as a guide for clinician researchers. The Declaration of Helsinki permits inclusion of vulnerable populations (e.g., children, mentally impaired individuals) as subjects of research and promotes a more paternalistic view of informed consent issues than does the Nuremberg Code. The Council for International Organizations of Medical Sciences in conjunction with the World Health organization also promulgated guidelines in 1982 with a recent revision in 1993, which included issues specific to research conducted in developing countries.[4] While these international guidelines have influenced the development of guidelines in the United States, they have no legal status in this country.

Researchers in the United States must comply with federal statutes, regulations, and guidelines that reflect the spirit of the international standards. The PHS Act[5] of 1985 established a requirement for institutional review boards to review research protocols involving human subjects. It also set out protections for these subjects and assigned the development of regulations, guidelines, and enforcement to the Department of Health and Human Services (DHHS). DHHS established the Office of Protection from Research Risks (OPRR) at NIH, which performs these functions. In 1991, regulations known as "The Common Rule" were promulgated and governed human experimentation. The Common Rule provided uniform guidelines for research funded by about 20 federal agencies and departments.[6]

All research protocols that involve human subjects must be evaluated by an independent review committee called an Institutional Review Board (IRB). The IRB has the responsibility to ensure and the authority to require that the protocols meet all of the other criteria listed next. The institution that the IRB serves cannot override IRB decisions. Criteria for approval of research as described in the regulations are summarized as follows:

1. Risks to subjects must be minimized by using sound research design, avoiding unnecessary risk, and using procedures already being performed on the subjects as part of their patient care.
2. Risks to subjects must be reasonable in relation to both anticipated benefits to the subject and also to the importance of any new knowledge that may be gained by the research.
3. Subject selection must be equitable considering both the purposes of the research and the setting in which it will be conducted. It must not be conducted in vulnerable populations such as children, prisoners, or pregnant women unless it meets additional strict criteria protecting those subjects.
4. Consent must be sought from every subject or, in some special cases that are carefully defined in the regulations, by the patient's legally authorized representative.
5. Consent must be documented.
6. Data must be monitored to ensure the safety of subjects.
7. Subjects' privacy and confidentiality must be maintained.

IV. INFORMED CONSENT

In medical settings the responsibility for obtaining consent from individuals for their participation in research protocols generally resides with the clinician members of research teams. Informed consent may, in some circumstances, be obtained by nonclinical personnel, particularly in psychological research, sociological research, or some biomedical research. Regardless of who is assigned responsibility for obtaining informed consent for a participation in an experimental protocol, each investigator who is participating in that research protocol is equally responsible for ensuring that proper informed consent procedures are being followed.

To obtain informed consent several criteria must be met. While a written form should never take the place of a careful discussion, most of the information discussed is normally included in a written consent form as well. Potential subjects must be told that they will be participating in an experimental protocol. Particularly when subjects are also patients, it is not enough for researchers to tell a them that the procedure is new, without telling them that the protocol itself is experimental. Patients must be given full disclosure of all information that might affect their decision to participate in the research. Such information includes the purpose of the experiment; the duration of the experiment; the risks that may be anticipated from the research and their probability; the severity of the potential risks; and the probability of any expected benefits. In addition, if the subject is also a patient, the subject must be advised of any alternative treatments that are available and how they compare to the proposed experimental protocol.

Full disclosure about risks also includes information about any financial liability that may be incurred by the subject as a result of participation in the protocol. Most medical insurance companies will not reimburse for experimental procedures, and there is growing concern about their willingness to pay for care for injuries suffered as a consequence of participating in research. Standard consent forms also include information about liability of the researchers or their institution for reimbursement for injury to subjects resulting from participation. Issues surrounding subject privacy and specific procedures for protecting subject confidentiality must be clearly spelled out. The subjects must be told whom to contact with problems and how those persons can be contacted.

Informed consent must be voluntary and noncoerced. While obvious coercion is clear to most researchers (excessive financial payments for participation, exaggerated promises of benefit, threats of loss of access to medical care), subtle coercion is sometimes more difficult to identify and to prevent. Groups of individuals who are likely to be vulnerable to coercion are offered special protections to safeguard their welfare. These groups include children, prisoners, pregnant women, mentally disabled persons, or economically or educationally disadvantaged persons. Other groups may share some of the characteristics of these vulnerable populations, such as terminally ill patients, active-duty military personnel, or individuals who are in a subordinate position to the researcher conducting the protocol.

A common example of a circumstance that has potential for coercion is participation of graduate students or postdoctoral fellows in research protocols conducted by the laboratories or the departments in which they work. Graduate students may be asked to do everything from donating blood to participating in highly risky procedures. Individual sacrifice for the common good of the lab is often central to the culture of the research environment. Thus, students and fellows who refuse to participate as subjects in an experimental protocol risk being viewed as uncooperative and disruptive. This perception can affect their relationships with other students and with laboratory directors and the future of their careers in science. Special care should be taken when asking for student volunteers to ensure both that confidentially is not compromised and that there is no coercion. Seeking participants from outside the laboratory effort may be more difficult in some respects, but is often the preferable way to recruit subjects.

For all research involving human experimentation, informed consent must be obtained from subjects. However, for some low-risk protocols, the Institutional Review Board may waive the requirement for a written, signed consent form documenting the discussion. Examples of such low-risk protocols might be interviews, surveys, chart reviews, and blood drawing. Even though the requirement for a written signed consent is waived, the researchers have the same responsibilities for conducting the research in an ethical and safe manner as they would for any protocol.

V. MOVING FROM BASIC RESEARCH TO RESEARCH ON HUMAN SUBJECTS

The move to bring an intriguing research idea from the bench to clinical trial requires great care. An interesting example can be found in a quote from Francis D. Moore written in 1969.

> At the present time, we are engaged in one of the largest mass human experiments of this type ever considered: the widespread use of oral contraceptives. It has been estimated that more than twenty-five million women have taken these tablets and that at any one time fifteen million women are taking them. Their effects after prolonged administration are entirely unknown. There is virtually no animal work reporting the continuous administration of these drugs for

more than three years, their impact—either on the psyche, ovaries, endometrium, or breast—on women after twenty years of continuous administration remains entirely unknown.[7]

As we know, regardless of the freedom that oral contraceptives have granted women, they have also caused a wide array of serious and not so serious complications. These include various cancers of the reproductive system, circulatory problems, and more minor conditions. It is only after 25 years of use that dosages and hormone combinations have been adjusted to reduce these risks.

In contrast to what occurred in the introduction of oral contraceptives, guidelines call for adequate experimental evidence to be present prior to undertaking human experimentation. Often this means many years of painstaking research both in experimental systems and in animal models. The original basic research may include research on animal models, but it may also mean in vitro studies on human cell lines or in human in vitro primary cultures. Animal and theoretical models of action for drugs or therapies are also useful.

Before undertaking clinical trials involving a new therapy in humans, adequate experimentation on large animals such as dogs, sheep, or primates may need to be performed. The large animal studies should mimic the human situation with regard to both disease conditions and therapeutic interventions. If, for example, a lengthy and complex surgery is to be done on patients with advanced disease who have weakened cardiovascular systems, then the animals operated on in the experimental trial should have similarly weakened physiology. This will provide evidence that individuals in a weakened condition can withstand the stress of the surgery. Clearly such models are not always possible, as when therapies for certain human diseases are being tried. Nevertheless, the condition of the animals used should mimic, as closely as possible, the conditions in proposed patients.

Experimentation with human subjects also requires that investigators have a thorough knowledge of the literature. They should be aware of other similar experiments that have been attempted, including those that have failed. They must be aware of alternative therapies: the other options for treatment of the condition under study. In fact, the informed consent for these patients must contain information on the alternatives. Investigators have an obligation to consider the physiology of the experiment as it involves not only the organ system or drug under study but the well-being of the entire individual. A drug with a history of causing serious circulatory problems in significant numbers of patients should not be undergoing clinical trial for use in mild cases of kidney disease.

Knowledge of the literature and data from preliminary studies will allow the investigator to balance the risks and the benefits of the experiment to the study subjects. The experimental design should ensure that the risks are not greater than the benefits obtained. Sometimes this becomes a very difficult balancing act, particularly in situations where the treatment considered or knowledge to be gained may cause multiple benefits to large numbers of people while putting a few people at serious risk. Some people still argue from this perspective that Milgram's experiments were ethical.

Data both from experimentation and from the literature should be incorporated into the design and execution of experiments on human subjects. Although all experimentation contains some amount of uncertainty, experiments should be designed to yield as few unexpected events as possible.

VI. EXPERIMENTAL DESIGN FOR RESEARCH ON HUMAN SUBJECTS

As with all research, research on human subjects should be based on a sound experimental hypothesis and sound experimental design. In contrast to other research, however, human experimentation using poorly designed experiments is itself unethical. This is because such research may cause harm to people at the same time as there is no hope that it will benefit either those subjects or others. A poorly designed experiment is not likely to yield usable scientific data and does not advance the scientific body of knowledge.

Valid design means that the questions asked must be experimentally valid and answerable. Further, the experiment designed must address the questions posed. The design must include estimates of the numbers of subjects such that statistical evaluation will in most cases yield significant differences when they occur and lack of significant difference when differences do not occur. The design must include adequate control groups. One caution with regard to control groups is that when studies are conducted on patients who are seriously ill, there cannot be a "no treatment" control group. The control group must be made up of patients who get a treatment that reflects the standard of care in the field. In this regard the experimental question cannot be, "What is the effect of this new treatment relative to no treatment?" The question instead must be, "What is the effect of the new treatment relative to the treatment options that are considered standard of care for that disease?"

Good experimental design also depends on having reasonable tools for measuring whatever parameter is being measured. If the measurement is a change in the subject's "quality of life," adequate, previously validated tools for measuring "quality of life" need to be found. The magnitude of the change must also be measurable with the tools available. If a change in the width of the endometrial stripe in the uterus is the endpoint, the instruments for measuring that change should be adequate to detect a change of the magnitude expected. These considerations are no different from those of other scientific experiments. But again, with human experimentation a failure to design the experiment properly is itself unethical.

Finally, protocols involving human subjects are required to contain criteria for stopping the experiment if the risks become too great. Protocols should have such stopping points built into them. Examples of such stopping points would be if as a result of the treatment or experiment subjects got clearly sicker, manifested severe psychological symptoms, or died in unexpected numbers.

VII. CONCLUSION

The design, construction, and execution of experiments on human subjects are the responsibility of the all members of a research team. Experimental design and execution must be based on sound experimental data and must be of the highest quality possible.

Notes

1. S. Milgram, *Obedience to Authority* (New York: Harper and Row, 1983); A. G. Miller, *The Obedience Experiments* (New York: Praeger, 1986).

2. "The Nuremberg Code," *Trials of War Criminals Before the Nuremberg Military Tribunals Under Control Council Law,* 10(2):181–82 (Washington, D.C.: U.S. Government Printing Office, 1949).

3. U.S. Department of Health and Human Services, "World Medical Association Declaration of Helsinki," in *Protecting Human Subjects: Institutional Review Board Guidebook,* A6-3 to A6-6 (Washington, D.C.: OPRR/NIH, 1993).

4. CIOMS/WHO, International Ethical Guidelines for Biomedical Research Involving Human Subjects (Geneva: CIOMS, 1993). For an interesting discussion of these guidelines see R. J. Levine, "New international ethical guidelines for research involving human subjects," *Annals of Internal Medicine* 119(4): 339–41. 1993.

5. Public Health Service Act as amended by the Health Research Extension Act of 1985, Public Law 99-158, Section 492, November 20, 1985.

6. Department of Health, Education, and Welfare, "The Belmont Report: Ethical principles and guidelines for the protection of human subjects of research," *OPRR Reports* (April 18, 1979) (Washington, D.C.: U.S. Government Printing Office, 1979).

7. F. D. Moore, "Therapeutic innovation: Ethical boundaries in the initial clinical trials of new drugs and surgical procedures," in *Experimentation with Human Subjects,* ed. Paul A. Freud (New York: George Braziller, 1969), pp. 358–78.

CASES FOR CONSIDERATION

CASE 1

A research lab was studying the biochemistry of white blood cells. Most of the work was performed by fellows, occasional graduate students, medical students on research rotations, technicians, and so forth. In the course of this work, samples of blood from unaffected individuals were needed as controls. It was common practice in this lab to draw blood from other individuals working in the lab to use as controls, and everyone took a turn in donating without any difficulty. No informed consent was ever obtained. Lab staff were generally aware of the kinds of experiments for which their blood was used and of the risks of having blood drawn. One member of the lab refused repeatedly to donate his blood because he was "anemic." Other lab personnel felt he was not being truthful and by refusing to donate was not a "team player." In fact, this lab member was in a high-risk group for HIV infection and had never been tested for the presence of HIV antibodies in his blood. He had no plans to be tested and was concerned about potential risks of exposure of his lab mates to HIV-infected blood if he should be positive. Rather than reveal his real reasons for not donating, he preferred to be considered uncooperative or uncollegial.

Questions for Discussion

1. Should informed consent have been obtained from these lab members before their blood was drawn?
2. Do you think the lab members may have felt coerced into donating?
3. Can confidentiality be protected in this setting?
4. Should graduate students be considered a vulnerable population for the purposes of human experimentation?
5. Do you think students applying to graduate school or for work in a particular lab should be told in advance of any ongoing human experimentation protocols conducted by that lab in which they might be asked to participate as subjects?
6. Would your responses change if the experiment involved significant discomfort or health risk to the lab members?

CASE 2

An active research protocol was ongoing at a major university to study endocrine function in tissues removed from hysterectomy patients. The approved protocol provided for informed consent to be obtained, for use of nonpathologic tissues to be removed during the scheduled surgery, and for a sample of blood to be taken at the time of surgery. One of the physicians involved with the study identified a unique patient who she felt would provide unusual and important information to the study. The patient was not undergoing hysterectomy but she had a rare tumor of the bowel that was known to be hormone responsive in many cases. The researchers agreed that their lab was tooled up to get a maximum of information from this tissue, which they felt could be of immense clinical value in future cases. The physician arranged with the researchers that when the surgery was done, three weeks hence, she would contribute the tissue to them. She spoke with the patient who agreed to the use of her tumor tissues and to contributing a tube of blood for the study protocol. She spoke with the pathologist, who agreed to have pathologic tissue, not needed for diagnosis, set aside for the research labs. The surgery was performed and the tissue removed and used for research.

Questions for Discussion

1. Did the physician and the basic researchers proceed appropriately in this situation?
2. What is the best protocol for dealing with unusual and unexpected situations that fall outside an approved human experimentation protocol?
3. Does the value of the data that may be obtained from this unusual patient affect the ethical considerations in this case?

CASE 3[1]

In 1990, the NIH Recombinant DNA Advisory Committee approved a clinical trial using gene-modified human cells in terminally ill cancer patients of a famous oncologist in collaboration with two colleagues, both well known for their expertise in gene therapy. Tumor-infiltrating lymphocytes (TIL cells) were extracted from cancer patients, genetically engineered to carry the gene for tumor necrosis factor (TNF), a cytokine that has toxic effects on tumors, and grown in culture to expand their numbers. The scientific premise of the experiment was that (1) TIL cells, when reinjected back into the patient, would localize to the tumor from which they had been derived; (2) these genetically modified TIL cells would secrete adequate levels of TNF to destroy the tumor. During the original approval process by the NIH RAC concern was expressed over lack of animal data demonstrating the soundness of the hypothesis. The investigators testified there was no suitable animal model. And regarding the hypothesis on which the trial was based, the data supporting localization of the TIL cells was open to interpretations other than the one offered by the investigators. Unknown to the NIH RAC, the investigators themselves also had concerns regarding expression of TNF by the modified TIL cells. One collaborator said later, "The first experiments were better than the later ones. In hindsight, the initial data were spotty. The initial promise they generated in the lab couldn't be maintained. It was really hard for us to get any expression of cytokines expressed in TILS."

After two years of clinical trials (nine patients), the study was routinely reviewed by the NCI Division of Cancer Treatment Board of Scientific Counselors in private and public hearings. The Board felt that the trials had not generated convincing data to show the experiment was working or ever would. It was not clear that the cells were localizing to tumor sites and they might be localizing in other organs, which they could potentially damage. In addition, the TIL cells were secreting levels of TNF well below those predicted to be effective in early experiments. The Board refused to renew funding for an outside contract lab to grow TIL cells and recommended the entire gene therapy portion of the trial be eliminated if no new evidence of feasibility was forthcoming.

NIH has no plans to halt the trial, and the principal investigator has other funding adequate to continue the study. In his defense of his project to the NCI Board, the investigator cited data from an unrandomized clinical trial that added cyclophosphamide (an anticancer drug) to the protocol and seemed to improve localization of the TIL cells to the tumor site. He also plans to switch to a new viral vector for carrying the TNF gene into the TIL cells, which he believes will increase TNF production by genetically modified TIL cells to 80% of the level predicted to be effective in tumor destruction. In addition, he said that all the appropriate committees had approved his protocol for experimentation in humans and he saw no reason to discontinue the trial.

Questions for Discussion

1. Did the investigators meet the criteria that should be satisfied before moving from the bench to human experimentation? If so, how? If not, why not?
2. How do you evaluate an ongoing trial? Should there be criteria for continuing as well as stopping a clinical trial?
3. Do the ethical responsibilities of the investigator change after his or her experiment/trial has been approved by oversight committees (e.g., NIH RAC) or an Institutional Review Board?
4. In terms of ethical responsibility, do the collaborators have less responsibility than the principal investigator?

Note

1. C. Anderson, "A speeding ticket for NIH's controversial cancer star," *Science* 259:1391–92. 1993.

The Nuremberg Code

1. The voluntary consent of the human subject is absolutely essential.

   This means that the person involved should have legal capacity to give consent; should be so situated as to be able to exercise free power of choice, without the intervention of any element of force, fraud, deceit, duress, overreaching, or other ulterior form of constraint or coercion; and should have sufficient knowledge and comprehension of the elements of the subject matter involved as to enable him to make an understanding and enlightened decision. This latter element requires that before the acceptance of an affirmative decision by the experimental subject there should be made known to him the nature, duration, and purpose of the experiment; the method and means by which it is to be conducted; all inconveniences and hazards reasonably to be expected; and the effects upon his health or person which may possibly come from his participation in the experiment.

   The duty and responsibility for ascertaining the quality of the consent rests upon each individual who initiates, directs or engages in the experiment. It is a personal duty and responsibility which may not be delegated to another with impunity.
2. The experiment should be such as to yield fruitful results for the good of society, unprocurable by other methods or means of study, and not random and unnecessary in nature.
3. The experiment should be so designed and based on the results of animal experimentation and a knowledge of the natural history of the disease or other problem under study that the anticipated results will justify the performance of the experiment.

Reprinted from *Trials of War Criminals Before the Nuremberg Military Tribunals Under Control Council Law* 2(10)(1949):181–82 by permission of the U.S. Government Printing Office, Washington, D.C.

4. The experiment should be so conducted as to avoid all unnecessary physical and mental suffering and injury.
5. No experiment should be conducted where there is an a priori reason to believe that death or disabling injury will occur; except, perhaps, in those experiments where the experimental physicians also serve as subjects.
6. The degree of risk to be taken should never exceed that determined by the humanitarian importance of the problem to be solved by the experiment.
7. Proper preparations should be made and adequate facilities provided to protect the experimental subject against even remote possibilities of injury, disability, or death.
8. The experiment should be conducted only by scientifically qualified persons. The highest degree of skill and care should be required through all stages of the experiment of those who conduct or engage in the experiment.
9. During the course of the experiment the human subject should be at liberty to bring the experiment to an end if he has reached the physical or mental state where continuation of the experiment seemed to him to be impossible.
10. During the course of the experiment the scientist in charge must be prepared to terminate the experiment at any stage, if he has probably [sic] cause to believe, in the exercise of the good faith, superior skill and careful judgement required of him that a continuation of the experiment is likely to result in injury, disability, or death to the experimental subject.

World Medical Association Declaration of Helsinki, 1989 Version

ADOPTED BY THE 18TH WORLD MEDICAL ASSEMBLY, HELSINKI, FINLAND, JUNE 1964, AND AMENDED BY THE 29TH WORLD MEDICAL ASSEMBLY, TOKYO, JAPAN, OCTOBER 1975, 35TH WORLD MEDICAL ASSEMBLY, VENICE, ITALY, OCTOBER 1983, AND THE 41ST WORLD MEDICAL ASSEMBLY, HONG KONG, SEPTEMBER 1989

INTRODUCTION

It is the mission of the physician to safeguard the heath of the people. His or her knowledge and conscience are dedicated to the fulfillment of this mission.

The Declaration of Geneva of the World Medical Assembly binds the physician with the words, "The health of my patient will be my first consideration," and the International Code of Medical Ethics declares that, "A physician shall act only in the patient's interest when providing medical care which might have the effect of weakening the physical and mental condition of the patient."

Reprinted from U.S. Department of Health and Human Services, *Protecting Human Subjects: Institutional Review Board Guidebook,* A6-3 to A6-6 (Washington, D.C.: OPRR/NIH, 1993) by permission of The World Medical Association, Inc., and with a warning from the WMA secretary-general dated 5 March 1996 that "there may be a small but fundamental change in the text of the Declaration introduced during our General Assembly in October, 1996. This will affect the text in Section II paragraph 3, which appears to some at present to exclude the use of placebo controlled trials."

The purpose of biomedical research involving human subjects must be to improve diagnostic, therapeutic and prophylactic procedures and the understanding of the aetiology and pathogenesis of disease.

In current medical practice most diagnostic, therapeutic or prophylactic procedures involve hazards. This applies especially to biomedical research.

Medical progress is based on research which ultimately must rest in part on experimentation involving human subjects.

In the field of biomedical research a fundamental distinction must be recognized between medical research in which the aim is essentially diagnostic or therapeutic for a patient, and medical research, the essential object of which is purely scientific and without implying direct diagnostic or therapeutic value to the person subjected to the research.

Special caution must be exercised in the conduct of research which may affect the environment, and the welfare of animals used for research must be respected.

Because it is essential that the results of laboratory experiments be applied to human beings to further scientific knowledge and to help suffering humanity, the World Medical Association has prepared the following recommendations as a guide to every physician in biomedical research involving human subjects. They should be kept under review in the future. It must be stressed that the standards as drafted are only a guide to physicians all over the world. Physicians are not relieved from criminal, civil and ethical responsibilities under the laws of their own countries.

I. BASIC PRINCIPLES

1. Biomedical research involving human subjects must conform to generally accepted scientific principles and should be based on adequately performed laboratory and animal experimentation and on a thorough knowledge of the scientific literature.
2. The design and performance of each experimental procedure involving human subjects should be clearly formulated in an experimental protocol which should be transmitted for consideration, comment and guidance to a specially appointed committee independent of the investigator and the sponsor provided that this independent committee is in conformity with the laws and regulations of the country in which the research experiment is performed.

3. Biomedical research involving human subjects should be conducted only by scientifically qualified persons and under the supervision of a clinically competent medical person. The responsibility for the human subject must always rest with a medically qualified person and never rest on the subject of the research, even though the subject has given his or her consent.
4. Biomedical research involving human subjects cannot legitimately be carried out unless the importance of the objective is in proportion to the inherent risk to the subject.
5. Every biomedical research project involving human subjects should be preceded by careful assessment of predictable risks in comparison with foreseeable benefits to the subject or to others. Concern for the interests of the subject must always prevail over the interests of science and society.
6. The right of the research subject to safeguard his or her integrity must always be respected. Every precaution should be taken to respect the privacy of the subject and to minimize the impact of the study on the subject's physical and mental integrity and on the personality of the subject.
7. Physicians should abstain from engaging in research projects involving human subjects unless they are satisfied that the hazards involved are believed to be predictable. Physicians should cease any investigation if the hazards are found to outweigh the potential benefits.
8. In publication of the results of his or her research, the physician is obliged to preserve the accuracy of the results. Reports of experimentation not in accordance with the principles laid down in this Declaration should not be accepted for publication.
9. In any research on human beings, each potential subject must be adequately informed of the aims, methods, anticipated benefits and potential hazards of the study and the discomfort it may entail. He or she should be informed that he or she is at liberty to abstain from participation in the study and that he or she is free to withdraw his or her consent to participation at any time. The physician should then obtain the subject's freely-given informed consent, preferably in writing.
10. When obtaining informed consent for the research project the physician should be particularly cautious if the subject is in a dependent relationship to him or her or may consent under duress. In that case the informed consent should be obtained by

a physician who is not engaged in the investigation and who is completely independent of this official relationship.

11. In case of legal incompetence, informed consent should be obtained from the legal guardian in accordance with national legislation. Where physical or mental incapacity makes it impossible to obtain informed consent, or when the subject is a minor, permission from the responsible relative replaces that of the subject in accordance with national legislation.

    Whenever the minor child is in fact able to give a consent, the minor's consent must be obtained in addition to the consent of the minor's legal guardian.

12. The research protocol should always contain a statement of the ethical considerations involved and should indicate that the principles enunciated in the present Declaration are complied with.

II. MEDICAL RESEARCH COMBINED WITH CLINICAL CARE (CLINICAL RESEARCH)

1. In the treatment of the sick person, the physician must be free to use a new diagnostic and therapeutic measure, if in his or her judgment it offers hope of saving life, reestablishing health or alleviating suffering.
2. The potential benefits, hazards and discomfort of a new method should be weighed against the advantages of the best current diagnostic and therapeutic methods.
3. In any medical study, every patient—including those of a control group, if any—should be assured of the best proven diagnostic and therapeutic method.
4. The refusal of the patient to participate in a study must never interfere with the physician patient relationship.
5. If the physician considers it essential not to obtain informed consent, the specific reasons for this proposal should be stated in the experimental protocol for transmission to the independent committee (I, 2).
6. The physician can combine medical research with professional care, the objective being the acquisition of new medical knowledge, only to the extent that medical research is justified by its potential diagnostic or therapeutic value for the patient.

III. NONTHERAPEUTIC BIOMEDICAL RESEARCH INVOLVING HUMAN SUBJECTS (NONCLINICAL BIOMEDICAL RESEARCH)

1. In the purely scientific application of medical research carried out on a human being, it is the duty of the physician to remain the protector of the life and health of that person on whom biomedical research is being carried out.
2. The subjects should be volunteers—either healthy persons or patients for whom the experimental design is not related to the patient's illness.
3. The investigator or the investigating team should discontinue the research if in his/her or their judgement it may, if continued, be harmful to the individual.
4. In research on man, the interest of science and society should never take precedence over considerations related to the well being of the subject.

Radiation: Balancing the Record

CHARLES C. MANN

Sensational news stories imply that dozens of unethical experiments placed unknowing people at risk. In reality, some studies were good, some were bad, and some were just ugly.

The nation's press is in the midst of one of its classic feeding frenzies—but this time scientists, rather than politicians or celebrities, are the main course. On November 15, the *Albuquerque Journal* published the first of three horrifyingly detailed articles about Manhattan Project scientists injecting plutonium into human beings. Three weeks later, Department of Energy secretary Hazel O'Leary told a press conference that she was "appalled, shocked, and saddened" by the report. The head of the agency—notorious in the past for secrecy—vowed to "open the archives," triggering a media firestorm. The ensuing tide of news reports, including cover stories in both *U.S. News & World Report* and *Newsweek,* flooded the nation with tales of evil scientists stuffing radioactive substances into prisoners, cancer patients, retarded children, even newborns.

Is it conceivable that just after the Second World War—when Nazi doctors tortured concentration camp inmates in the name of science—

U.S. researchers treated unknowing patients with similar disregard? Or, as some researchers have claimed in defense, did the experiments have a historic context and scientific value that provides justification for these seemingly inhumane actions?

Press reports have overwhelmingly favored the former, cynical, explanation. The *Albuquerque Journal* described a single incident in which 18 terminal patients were injected with plutonium. Then newspapers such as *The Boston Globe* and *The Portland Oregonian* dug up research that involved feeding radioactive milk to retarded children and zapping the testicles of prisoners with high-energy radiation. Other journalists added to the impression of widespread abuse with their belated discovery of a 1986 congressional report detailing cases in which "nuclear guinea pigs" tested radioactive compounds such as tritium and technetium.

But an inquiry by *Science* shows that reality is, as usual, more complex than this sensational picture. For instance, at least five of the 31 experiments in the congressional report were apparently performed by researchers on themselves—hardly unknowing human guinea pigs. At least nine others cited in that document (known as the Markey Report, after Edward Markey (D–MA), head of the subcommittee that produced it) involved the use of minute, harmless quantities of radioactive isotopes to follow biochemical reactions within the body. "The trouble with the reporting today is that it doesn't make any distinctions," complains Sanford Miller, dean of the Graduate School of Biomedical Sciences at the University of Texas Health Science Center in San Antonio. "It's all radiation with a capital R. But there's various radiations—it's not a single golem rising out of the grave. And how people have thought about it over time is a lot more complicated than the newspapers make out."

That doesn't mean all of this research was valid—or even defensible. All evaluations are provisional, because some of the researchers are dead and complete explanations for their behavior are hard to come by, and records and accusations are still coming in. But at this point, it appears that the radiation experiments in the United States can be broadly classified in three groups. In one, researchers knowingly inflicted potential harm on patients, using methods that are difficult to justify even by the standards of the past. By contrast, a second, larger group of investigations involved perfectly good work by any standards, with appropriate safeguards taken. And a third group of studies falls between these extremes: The experiments provided useful information but had ethical flaws.

Doing Possible Damage

Most of the media attention has been paid to the injections of plutonium that took place between April 1945 and July 1947—a period that began a few months before Hiroshima and ended at the time of the Nuremberg trials. The nation then faced a serious public health problem, recalls J. Newell Stannard, a health physicist at the University of California, San Diego, and author of the 2000-page study *Radioactivity and Health: A History*. "Thousands of workers at the Manhattan Project had been potentially exposed to plutonium," he says. "Physicians were able to monitor how much [plutonium] the workers excreted, but they didn't know how much they had taken in, because the exposures were accidental."

Since it already seemed clear that many more workers would be exposed to radioactivity over time, plant safety officials were frantic for information about the effects of plutonium. But nobody even knew whether it was quickly excreted, limiting its potential for danger, or retained in the body, where it could keep irradiating tissue for years. A few studies had been done with rats, mice, rabbits, and dogs. But the data were contradictory, partly because different species metabolize plutonium differently. So radiologist Stafford Warren and the other members of the Manhattan Project Health Group came up with the plan of introducing known quantities of plutonium into the bodies of terminally ill volunteers. Their already-short life expectancies would both duck the question of long-term harm and allow any remaining plutonium in their bodies to be measured at an autopsy. Warren's team chose 18 men and women, all terminal patients, from San Francisco, Chicago, and Rochester.

Today, any research like this must obtain human subjects' "informed consent." Such consent means that the subjects of the experiment appreciate the known—and suspected—risks of participation and voluntarily agree to participate even after knowing those risks. At the time of Warren's plan, the term "informed consent" was not yet widely used, but the principle had been established in court cases during the 1930s, and other researchers did follow it. The idea was explicitly codified during the prosecution of the Nazi doctors, which began in December 1946.

For plutonium, though, informed consent was out of the question—because the military wouldn't hear of it. General Leslie Groves, director of the Manhattan Project, was "paranoid about security," says Stannard.

"Plutonium couldn't even be named—it had to be called 'product.' All [Warren and the other doctors] could tell them was that they were going to get a product in a small dose."

Even taking the rigid constraints imposed by the military into account, there is little to suggest that the subjects were thoughtfully chosen—a necessity if one has a small number, says Richard Griesemer, deputy director of the National Institute of Environmental Health Sciences (NIEHS). According to the *Albuquerque Tribune,* at least six had been wrongly diagnosed and were not about to die; two more were suffering from conditions disrupting the metabolic pathways the investigators were examining. Many injectees were apparently lost to follow-up. Only five are known to have been autopsied, one of the express purposes of the research.

According to toxicologists such as James Huff, a senior scientist in the environmental carcinogenesis program at NIEHS, researchers usually minimize possible toxic effects by administering slowly increasing doses to subjects. This did not occur. The first three doses were .29, .04, and 3.54 microcuries, respectively. The second dose was the lowest given; the third, administered less than a month later, was the third highest.

"The experiment did not have a rigid protocol established by some central authority," says biophysicist Patricia Durbin from Lawrence Berkeley Laboratory, who worked with the plutonium data for decades. Nonetheless, she says, the study was invaluable. "It is the *only* human data where the actual quantity inside the body is known and the time it was acquired is known." And Kenneth L. Mossman, president of the Health Physics Society, told Congress on 18 January that data from this "extremely important" study "serves as the principal database for current plutonium standards."

But even if the results stand up, many scientists say, the ethics do not. "People didn't know a lot of the things we know now," says Huff. "Toxicology didn't really exist as a field. Still, they knew that you shouldn't give people things that might harm them in the long term and try to get around it by saying they would die soon, anyway."

Two other sets of experiments raise similar concerns. Between 1963 and 1971, Carl G. Heller of the University of Oregon and the Pacific Northwest Research Foundation exposed the testicles of 67 prisoners at Oregon State Prison to ionizing radiation. One of Heller's protégés, C. Alvin Paulsen of the University of Washington, irradiated the testicles of 64 inmates at Washington State Prison between 1963 and 1970.

The reason for the testicular work was a 1962 accident at the Hanford Nuclear Reservation, in Richland, Washington, which exposed three workers to high doses of gamma radiation. Hanford officials asked Paulsen, a reproductive physiologist, to inform the men about their prospects for fatherhood. Little was known, it turned out. This was alarming to contemplate in an era that envisioned the rapid spread of nuclear power.

"We didn't have the knowledge for effective safety standards," Paulsen says. "I decided it was important to have certain information, such as the ED-50—the effective dose that would impair sperm production in 50% of men." Some animal research had been done, but, as in other areas of radiation research, the behavior of animal and human reproductive systems often differs. The best way to learn more, Paulsen reasoned, was to expose men to single blasts of radiation and measure the reaction, gradually increasing the amount with each group of subjects to construct a dose-response curve. "And that," he says, "brought up the issue of what type of population should be exposed."

Experimentation with prisoners was not unusual at the time—Heller, Paulsen's mentor, had been using them for years. "You had a wonderfully controlled population that was highly cooperative," dryly observes Wil Nelp, chairman of the department of nuclear medicine at the University of Washington. "They couldn't go anywhere, so follow-up was easy." Inmates often were housed in special state-hospital wards and fed fancy diets. Researchers sometimes filed notices of cooperation in their records. These were hard things for prisoners to turn down. "It was a good deal for them," Nelp says. "Probably too good of a deal."

Paulsen obtained permission for his study from the university, the state, and the prison. He then asked for volunteers among the inmates, asking them to promise to have vasectomies afterward. Before starting, Paulsen says, he privately interviewed each volunteer "and gave him every opportunity to say yea or nay." Then he gave all the participants (except a control group) between 7.5 to 400 roentgens, a high dose. Heller began a similar program a month later, with still higher doses: 8 to 600 roentgens.

Despite the chance to "say yea or nay," questions soon arose about the nature of the prisoners' consent. In 1966, the late anesthesiologist Henry Beecher published two landmark articles in the *New England Journal of Medicine* arguing, among other things, that patients' consent to a procedure that will harm them indicates some coercion, since people who are free to choose won't usually allow themselves to be hurt.

Further extension of this reasoning implies that prisoners cannot give informed consent at all, since their circumstances make them particularly prone to such coercion. The articles provoked enormous controversy, and, concerned about the combination of radiation and prisoners, the University of Washington halted Paulsen's work in 1969, rejecting his pleas to continue with additional measurements. Heller had a stroke in 1972, ending his efforts before anyone else could.

Yet the ethical flaws did not obviate its scientific interest. A 1974 paper based on Heller's prisoner work has been cited 135 times, according to the Institute for Scientific Information, a rate that places it in the top 1% of all the papers they track. And Paulsen says, "My colleagues were interested in what I was doing. I was not off by myself."

Valid Research

In sharp contrast to these experiments, much of the other work attacked in the Markey Report seems entirely blameless. Five 1950s-era experiments, for example, involved volunteers bathing patches of their skin in tritium, a radioactive form of hydrogen gas. One subject was exposed over his entire body. These people, the report charges, "thus became nuclear calibration devices." True. But the saving grace, which the report doesn't mention, is that in at least five of these studies the calibration devices apparently were the experimenters themselves.

At the time, atmospheric nuclear tests were creating lots of tritium, and scientists like Harry A. Kornberg of Hanford wanted to find out whether people could absorb tritium from water and air. After exposing rats, the team learned that living systems could, in fact, absorb tritium. But, again, the key question was whether the rat data could be applied to people.

The only way to find out was to perform tests on human beings. According to Chester W. DeLong, a team member who is now retired in Virginia, the volunteers were other researchers at the Hanford health lab. Kornberg insisted on doing the whole-body immersion himself. "He said, 'Well, I'm past child-bearing stage and I can tolerate it,'" recalls DeLong. "And I don't want anybody else exposed this way.'" And the work paid off, showing that people absorbed tritium four times faster than rats.

These results became mostly irrelevant, however, after the atmospheric test ban treaty was signed in 1963. But such self-experimentation was and is considered acceptable, and even heroic, according to *Who*

Goes First?, Lawrence Altman's 1987 book on the subject. As for the "guinea pig" characterization in the Markey Report, DeLong says, "They never bothered to call and ask me about the work."

Four other experiments involving radioactive tracers that are mentioned in the Markey Report seem to be equally acceptable ethically. In those experiments, researchers introduced tiny amounts of short-lived radioactive isotopes into human subjects. The radioactivity allows the tracers to be followed through the body. Afterward, they decay into harmless substances. In some media reports, this has been described as feeding people radioactivity. True enough—but the technique has been used for 60 years with no apparent ill effect.

In one example from 1965, the University of Washington's Nelp and two colleagues from the Pacific Northwest Laboratory, then a nonprofit organization based in Richland, Washington, injected technetium-95 or -96 into eight students and housewives in Seattle. Because technetium has a half-life of just a few hours, doctors hoped they would be able to inject it into patients, use its radiation to take a kind of internal x-ray, and then have it decay quickly into a harmless substance.

"I said that the first thing we ought to do was find out the ABCs of how technetium behaved in the body," Nelp says. He found eight volunteers—students and housewives—by word of mouth. After he explained the procedure to them, the volunteers signed consent forms. They stayed in or visited hospitals for up to 2 months while researchers collected samples of their blood, tears, perspiration, urine, and feces.

Partly as a result of this work, which revealed that the body quickly excretes technetium, it is now widely used in nuclear medicine. Doctors attach it to phosphate compounds and inject the ensemble into patients and wait for the body to incorporate the phosphates into bone, along with the technetium. "Afterward, you take a total body survey," Nelp explains. "You can see every bone in the body with much less radiation than if you took a series of x-rays."

So why was Nelp's research singled out in the Markey Report? Apparently because the dose given to the volunteers—20 to 60 microcuries—was up to six times higher than what the report described as the "occupational maximum permissible body burden" for these isotopes. But a "body burden" refers to the long-term buildup of a substance within the body, not a one-time shot, as in Nelp's experiment. And "you can't translate a one-time exposure for a volunteer into a standard designed for workers who might be breathing in something 8 hours a day, five days a week," says Lauriston S. Taylor, former head of the National Council on Radiation Protection and Measurements, the independent advisory panel that has recommended U.S. radiation safety standards since 1930. The difference, in other words, is the difference between a heart patient's one-time binge on six hamburgers and a diet of one greasy hamburger a day, which is more harmful.

Much of the research now being denounced in the media was tracer research, Taylor says. "It's not generally appreciated, apparently, that the magnitude of the dose from these tracer tests is just awfully small.

The Gray Area

In between the reprehensible and the praiseworthy are studies that appear to have been designed to ensure no harm came to people, but fell short on informed consent. Particularly worrisome is research on disabled or unconscious people—a red flag to medical ethicists today. Perhaps the best-known cases occurred in 1954 and 1956, when scientists affiliated with the Massachusetts Institute of Technology (MIT) Radioactivity Center fed radio-tagged milk to 36 mentally-retarded children at the Fernald School in Waltham, Massachusetts. The purpose was to answer a then-current puzzle: whether children who eat oat cereals, which are rich in compounds that bind to calcium, are thereby flushing calcium through their systems before their bones can use it to grow.

Like Nelp, the MIT researchers used a radioactive tracer—in this case, calcium-32. The idea was to "label" children's milk with this isotope and find out whether eating oat cereal would affect the amount that stayed in the body. This involved feeding children a uniform diet and collecting all their urine and feces for some time—a complicated prospect. Team leader Robert Harris had decided that such experiments would best succeed if the subjects were in a confined location and under medical supervision. The Fernald children met those criteria. The experiment suggested that oatmeal did indeed flush calcium from the system, but at a slow rate that would only affect children with very low-calcium diets.

Only "a tiny, tiny amount" of radioactive calcium was used, says Constantine Maletskos, a member of the team. According to MIT Radiation Protection Office director Francis Masse, the dose was 4 to 11 millirems above background. (Typical background levels are about 300 millirems.) By comparison, a typical treatment for hyperthyroidism involves hitting the thyroid with a drink that delivers about 10 million millirems. "They would have had more if they had flown to Denver for a while," Maletskos says, where they would have been exposed to that high-altitude city's greater number of cosmic rays.

Although the doses of radiation were small, the consent for the experiment would not have met today's standards. "In those days doctors were the kings of their facilities," say Maletskos. "They were in charge of their patients. [The Fernald supervisors] told us they had consent, and it would never have occurred to us to question them." Maletskos says he was horrified to learn on 26 December in a story from *The Boston Globe* that the consent forms sent to the parents by the school had neglected to mention "radioactivity." The school merely asked parents about participating in nutritional experiments. But even if the forms had mentioned radioactivity, there are doubts consent could ever be properly obtained from retarded subjects or their parents. Indeed, today the whole issue of informed consent by the mentally impaired is regarded as so blurred that experimenters believe they should not be used as a study population.

Similar questions of consent dog some of the cases mentioned in the Markey Report. An example is the injection of radioactive uranium-235 into at least 11 comatose, terminal cancer patients between 1953 and 1957 by William Sweet of Massachusetts General Hospital, in Boston, and his associates. The procedures were done as part of the development of what is called "neutron-capture therapy." Neutron-capture therapy takes advantage of the fact that tumors absorb more of certain isotopes than healthy tissues do. After placing those isotopes in the body, doctors bombard the patients with neutrons, which split the isotopes, releasing radiation that kills surrounding cancer cells.

In the 1950s, this idea was little more than plausible-sounding speculation. No one knew which isotope would best be absorbed by tumors. Sweet decided to find out. After obtaining permission for the injections from the patients' families, he carried out the study. The results were disappointing. Uranium, it seemed, was not absorbed in sufficient quantities by the tumor to make the therapy practical; in current attempts at neutron-capture therapy, boron is used.

Even at the time this work could have aroused qualms. In 1953, the year Sweet began his experiments, the British Medical Council campaigned against the use of comatose subjects in research. And as far back as 1948, the Federation of American Societies of Experimental Biology expressed concern that experimenting on the "hopelessly incurable" would "corrupt" the doctor-patient relationship, because it could make their rapid deaths desirable if an autopsy was needed. Nowadays, research with no potential for direct benefit to the terminally ill subject is generally avoided

Yet these matters of consent and safety frequently fall into gray areas, as researchers acknowledge. People with AIDS, for instance, clamor to be experimented on with medications whose effects are so poorly understood that neither physician nor patient can give consent truly informed by knowledge of risks and benefits. "Who knows what people will think of that in the future?" Stannard says. "We should be humble and wonder what we now are doing that will horrify our descendants." Unlike radioactive decay rates, the rate of change in morality standards has never been accurately measured.

Contributors

Robert Arzbaecher, PhD, is Professor and Director of the Pritzker Institute of Medical Engineering at Illinois Institute of Technology. He also has the rank of Professor in the Department of Internal Medicine, University of Chicago. Prior to his current appointment in 1981 he was Chairman of the Department of Electrical & Computer Engineering at University of Iowa. He has written over one hundred research publications in the area of engineering approaches to problems in clinical cardiology.

Edward M. Berger, PhD, is Professor of Biology and Dean of Graduate Studies at Dartmouth College. He also serves as co-director of the Molecular Genetics Center at Dartmouth. Berger is author or co-author of more than eighty research articles on genetics and molecular biology. He was co-principal investigator on an NIH grant to study ethical concerns raised by the Human Genome Project. In 1990 he was elected membership into the Human Genome Organization.

Stephanie J. Bird, PhD, is Special Assistant to the Provost of the Massachusetts Institute of Technology, where she assists departments and research groups in the development of educational programs that address ethical issues in research practice and the professional responsibilities of scientists. She is a laboratory-trained neuroscientist whose research interests now focus on the ethical, legal, and social policy implications of scientific research. She has written numerous articles on predictive testing for hereditary disease, and on mentoring and other responsibilities of science professionals. She is an editor of the journal *Science and Engineering Ethics.*

Marilyn J. Brown, DVM, MS, is the Director of the Dartmouth College Animal Care and Use Program and an Assistant Professor of Microbiology at Dartmouth Medical School. She has involvement in numerous professional organizations such as the American Association for Laboratory Animal Science and the Association for Assessment and Accreditation of Laboratory Animal Care, and has served on the Boards of Directors of the American Society of Laboratory Animal Practitioners, and the American College of Laboratory Animal Medicine. She assisted on the national panel that revised the *Guide for the Care and Use of Laboratory Animals,* and her publications include several chapters in the National Agriculture Library's publication *Essentials for Animal Research.*

Jeffrey L. Doering, PhD, is Professor of Biology at Loyola University Chicago and a former fellow of the university's Center for Ethics. His area of research is human molecular genetics.

Deni Elliott, EdD, is University Professor of Ethics and Director of the Practical Ethics Center at the University of Montana. She has co-produced three videodocumentaries and edited two other books in applied and professional ethics.

Bernard Gert, PhD, is Eunice and Julian Cohen Professor for the Study of Ethics and Human Values at Dartmouth College and Adjunct Professor of Psychiatry at Dartmouth Medical School. His Books Include *Morality: A New Justification of the Moral Rules, Philosophy in Medicine: Conceptual and Ethical Issues in Medicine and Psychiatry,* and *Morality and the New Genetics.*

David E. Housman, PhD, is Ciba-Geigy Professor of Biology at the Massachusetts Institute of Technology's Center for Cancer Research. He is involved in research on the molecular basis of human genetic disease, including familial cancer, neurodegenerative and neuromuscular diseases, and psychiatric disorders. Dr. Housman is a member of the National Academy of Sciences and the American Academy of Microbiology.

Karen J. Lomax, MD, a Pediatrician, is the Director of the National Center for Clinical Ethics (NCCE) of the Department of Veterans Affairs and Chairperson of the VA National Headquarters Bioethics Committee. She has responsibility for overseeing all aspects of NCCE's nationwide programmatic ethics activities and has been especially active in healthcare policy development for the Veterans Hospital Administration. She was previously Head of the Gene Therapy Unit Laboratory of Clinical Investigation at the National Institute of Allergy and Infectious Disease, NIH, where her chief research interest was in the molecular genetic basis of phagocytic disorders.

Allan U. Munck, PhD, is the Third Century Professor of Physiology at Dartmouth Medical School. He holds an MS degree in Chemical Engineering and a PhD in Biophysics. He joined the Department of Physiology of Dartmouth Medical School in 1959 as an Assistant Professor and has been Professor of Physiology since 1967. His research focuses on the molecular and physiological actions of glucocorticoids and on the mechanisms of their effects in lymphoid and other tissues. A recent area of interest is the relation of the therapeutic applications of glucocorticoids, previously referred to as pharmacological actions, to their physiological functions.

Judy E. Stern, PhD, is an Associate Professor of Obstetrics and Gynecology at the Dartmouth-Hitchcock Medical Center, where she is Professional Director of the Human Embryology and Andrology Laboratory. Her research interests are in the fields of reproductive immunology and infertility. She is an active member of the Institute for Applied and Professional Ethics at Dartmouth College, where she has participated in projects on scientific research ethics and on the ethical issues raised by assisted reproductive technologies.

Judith P. Swazey, PhD, is President of The Acadia Institute in Bar Harbor, Maine, an independent nonprofit center for the study of issues concerning medicine, science, and society that she founded in 1984. She also is an Adjunct Professor of Social and Behavioral Sciences at Boston University Schools of Medicine and Public Health. Her research, writing, and teaching have focused on social, ethical, legal and policy issues in biomedical research and health care; professional ethics; and graduate and professional education. She has written or edited thirteen books and numerous articles dealing with topics in these fields.

Vivian Weil, PhD, is Director of the Center for the Study of Ethics in the Professions and Professor of Ethics at the Illinois Institute of Technology. She has written widely on engineering ethics and scientific research ethics and co-edited the volume, *Owning Scientific and Technical Information: Value and Ethical Issues.*

Patricia H. Werhane, PhD, is the Ruffin Professor of Business Ethics in the Darden School at the University of Virginia. Her works include *Ethical Issues in Business,* cited with Tom Donaldson (4 editions), *Persons, Rights, and Corporations; Adam Smith and His Legacy for Modern Capitalism;* and *Skepticism, Rules, and Private Languages.* She is Editor-in-Chief of *Business Ethics Quarterly,* and on the editorial boards of *American Philosophical Quarterly, Journal of Value Inquiry,* and *Public Affairs Quarterly.*

Library of Congress Cataloging-in-Publication Data

Research ethics : a reader / edited by Deni Elliott and Judy E. Stern.

p. cm.

ISBN 0–87451–797–4 (pbk. : alk. paper)

1. Research—Moral and ethical aspects. 2. Research—Moral and ethical aspects—Study and teaching. I. Elliott, Deni. II. Stern, Judy E. III. Institute for the Study of Applied and Professional Ethics.

Q180.55.M67R44 1997

174'.95—dc20 96–34654